Oceanography

Earth Science Lab Manual

Second Edition

Shannon Othus-Gault

Chemeketa Press | Salem, Oregon

Oceanography: Earth Science Lab Manual
© 2025 Shannon Othus-Gault.

ISBN-13: 978-1-955499-40-8

All rights reserved. Edition 1 2019. Edition 2 2025.

Chemeketa Press
Chemeketa Community College
4000 Lancaster Dr NE
Salem, Oregon 97305
collegepress@chemeketa.edu
chemeketapress.org

Cover design by Ronald Cox
Interior design by Ronald Cox

References to website URLs were accurate at the time of writing. Neither the author nor Chemeketa Press is responsible for URLs that have changed or expired since the manuscript was prepared.

Printed in the United States of America.

Land Acknowledgment
Chemeketa Press is located on the land of the Kalapuya, who today are represented by the Confederated Tribes of the Grand Ronde and the Confederated Tribes of the Siletz Indians, whose relationship with this land continues to this day. We offer gratitude for the land itself, for those who have stewarded it for generations, and for the opportunity to study, learn, work, and be in community on this land. We acknowledge that our College's history, like many others, is fundamentally tied to the first colonial developments in the Willamette Valley in Oregon. Finally, we respectfully acknowledge and honor past, present, and future Indigenous students of Chemeketa Community College.

Contents

1 The Ocean Floor

Purpose

To learn to use latitude and longitude, the grid system found on maps, to find various locations on maps.

Materials

- ❑ Ocean Floor Map
- ❑ Calculator
- ❑ Age of Oceanic Lithosphere Map
- ❑ Political or Geographic World Map

Part 1: Latitude and Longitude

Q1. Find the latitude and longitude of the following islands. Report the island's locations using degrees (for now, ignore minutes and seconds). Also, make sure you are using the map provided to find these locations. Do not use your smartphone (I can generally tell if you use your phone by how accurate you are).

Example: Sumatra, Indonesia: 0°S, 101°E

    a.   New Guinea:_____

    b.   Ireland: _____

    c.   Madagascar: _____

    d.   Luzon, Philippines: _____

    e.   Sicily, Italy: _____

    f.   Hawaii: _____

    g.   Iceland: _____

    h.   Sri Lanka: _____

    i.   Puerto Rico: _____

    j.   Galapagos Islands: _____

Q2. Which of these islands is closest to the equator? _____

Q3. Which of these islands is closest to the poles? _____

Part 2: Time Zones

Time zones are determined by longitude. Because the globe is a sphere, it represents 360°. Also, there are 24 hours in a day, and all 24 hours are represented somewhere on the globe by time zones. 360° can be divided by 24 hours, leaving us with 15° of longitude for each hour or time zone. This means that if we know the difference of longitude between cities, we can calculate their time difference (ignore any numbers after the decimal).

For example, the time difference between McMinnville (45°N, 123°W) and New York (40°N, 74°W) is:

123° – 74° = 49° – McMinnville and New York are 49° apart.
49° / 15° = 3.2 hours – McMinnville and New York are 3 hours or time zones apart.

Q4. What is the time difference between Ireland and Hawaii (use Part 1)? Show your work.

Q5. Why are time zones based on longitude? Hint: think about why the sun sets in the west.

Part 3: Calculating Nautical Distance

Nautical miles are used in the ocean because they are based on the circumference of the Earth. One nautical mi (1nm) is equal to one minute (1') of latitude. Since there are 60' in 1° of latitude, then 1° of latitude is 60nm. You can use this conversion (60nm = 1°) to calculate the distance of various locations on Earth and can also convert oceanic nautical miles to statute miles and kilometers, measurements used on land using the formulas below.

For example, the distance between New York (40.7°N, 74°W) and Miami (25.8°N, 80.1°W) is:

40.7° – 25.8° = 14.9° — New York and Miami are 14.9° apart.
14.9° * 60 nm = 894 nm — New York and Miami are 894 nautical miles apart.

Q6. How far apart are McMinnville, OR (45°12'N, 123°12'55W) and San Francisco, CA (37°46'N, 122°25'W) in Nautical Miles? You do need to use the minutes here. Convert the minutes to degrees by dividing by 60 minutes. Show your work.

Q7. How far is that in statute miles (1nm = 1.15mi)? Show your work.

Q8. How far is that in kilometers (1mi = 1.609km)? Show your work.

Part 4: Oceanic Crust vs. Continental Crust

Examine the Age of Oceanic Lithosphere Map to determine if the following locations are oceanic or continental crust. *Note: continental crust can be covered with ocean water. Those areas are gray in color on your map.* Check the box of the correct crust type for each area below.

Q9. What type of crust underlies the western Florida Coast?

- ❑ Continental
- ❑ Oceanic

Q10. What type of crust underlies the Bering Strait?

- ❑ Continental
- ❑ Oceanic

Q11. What type of crust underlies the Gulf of California?

- ❑ Continental
- ❑ Oceanic

Q12. What type of crust underlies the Red Sea?

- ❑ Continental
- ❑ Oceanic

Q13. What type of crust underlies the North Sea?

- ❑ Continental
- ❑ Oceanic

Q14. What type of crust underlies the Mediterranean Sea?

- ❑ Continental
- ❑ Oceanic

Q15. What type of crust underlies Australia?

- ❑ Continental
- ❑ Oceanic

Q16. What type of crust underlies the Hawaiian Islands?

- ❑ Continental
- ❑ Oceanic

Q17. What type of crust lies between Madagascar and Africa?

- ❑ Continental
- ❑ Oceanic

Q18. What type of crust lies between China and Japan?

- ❑ Continental
- ❑ Oceanic

Q19. What type of crust lies between Australia and Papua New Guinea?

- ❑ Continental
- ❑ Oceanic

Q20. What type of crust is Indonesia?

- ❑ Continental
- ❑ Oceanic

Q21. What type of crust is on the northern coast of Iceland?

- ❑ Continental
- ❑ Oceanic

Part 5: Locations

Clearly label of all of the following features below on the included World Map (figure 1.1).

❑ **Continents:** Label all 7 continents.

❑ **Oceans:** Label all 5 oceans.

❑ **Islands:** Label the following major islands:
 ❑ Aleutians
 ❑ Galapagos
 ❑ Greenland
 ❑ Hawaii
 ❑ Iceland
 ❑ Japan
 ❑ Java
 ❑ New Zealand
 ❑ Philippines
 ❑ Sumatra

❑ **Mountain Ranges:** Label the following major mountain ranges by writing along the feature:
 ❑ Alps
 ❑ Andes
 ❑ Cascades
 ❑ Himalayas
 ❑ Zagros
 ❑ Appalachian

❑ **Major Map Features:** Draw the following features:
 ❑ Equator
 ❑ Prime Meridian
 ❑ Compass rose

Figure 1.1

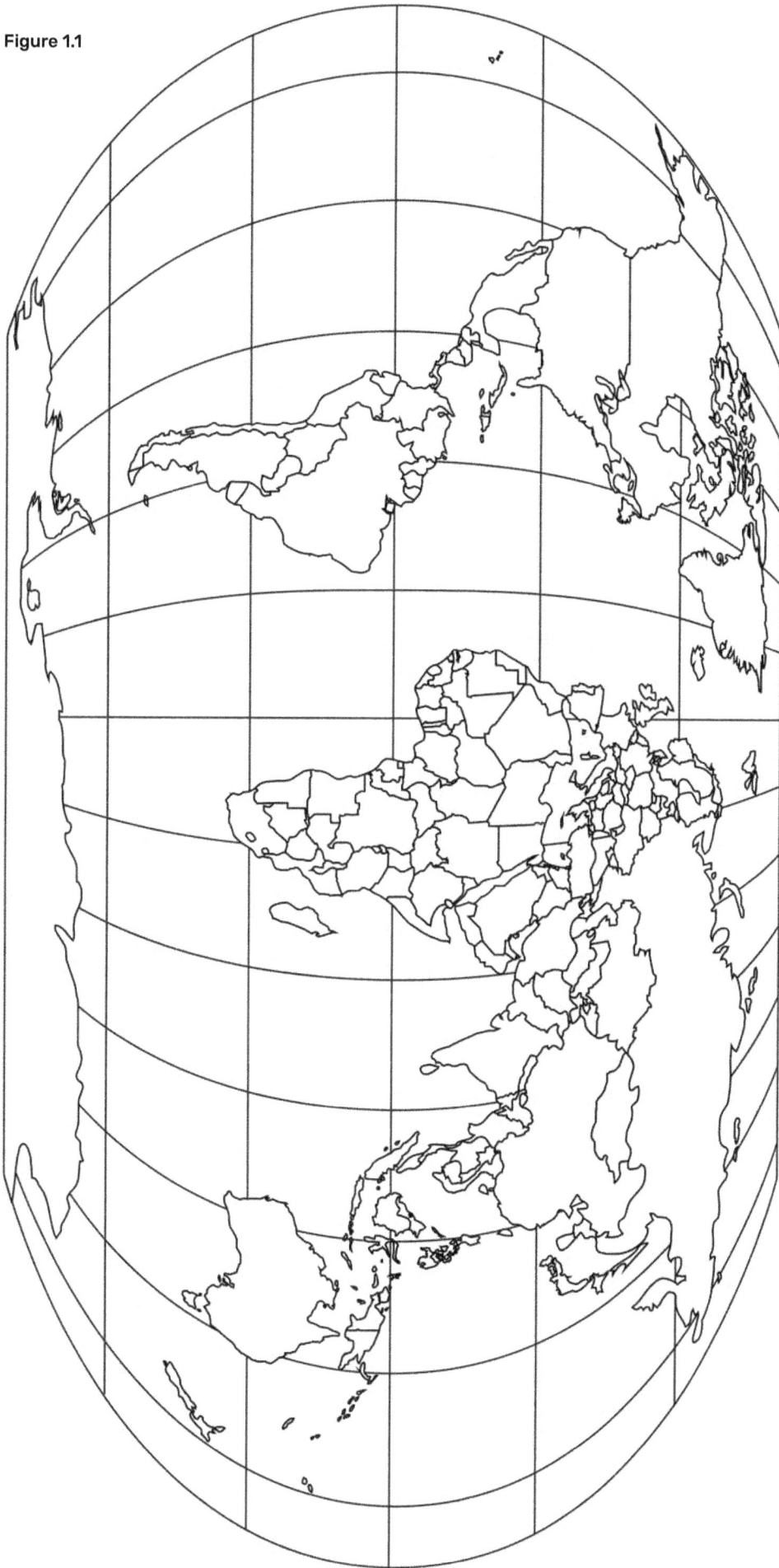

Name _____ Section _____ Date _____

Group Members _____

2 | Plate Tectonics

Purpose

To locate topographic features in the ocean, introduce plate tectonics, calculate plate motion using both a hotspot track and by calculating spreading rate, and to differentiate the densities of Earth's crust.

Materials

- ❏ Volcanology Map
- ❏ Seismology Map
- ❏ Plate Boundary Map
- ❏ Calculator
- ❏ Ruler

- ❏ Graph paper
- ❏ Rock sample of basalt
- ❏ Rock sample of granite
- ❏ Graduated cylinder
- ❏ Scale

Part 1: Plates, Volcanoes, and Earthquakes

For the following set of questions, use the maps provided (figure 2.1, 2.2, and 2.3) and answer the questions about how plate tectonics relate to features and events that occur in the ocean.

Q1. Using the earthquake map, find an area in the oceans where there is a linear occurrence of shallow earthquakes (red dots). Describe the location of the area.

Figure 2.1

SCIENTIFIC SPECIALTY: VOLCANOLOGY

Red dots indicate currently or historically active volcanic features

This list obtained from the Smithsonian Institution

This map is part of "Discovering Plate Boundaries," a classroom
exercise developed by Dale S. Sawyer at Rice University (dale@rice.edu).
Additional information about this exercise can be found at
http://terra.rice.edu/plateboundary

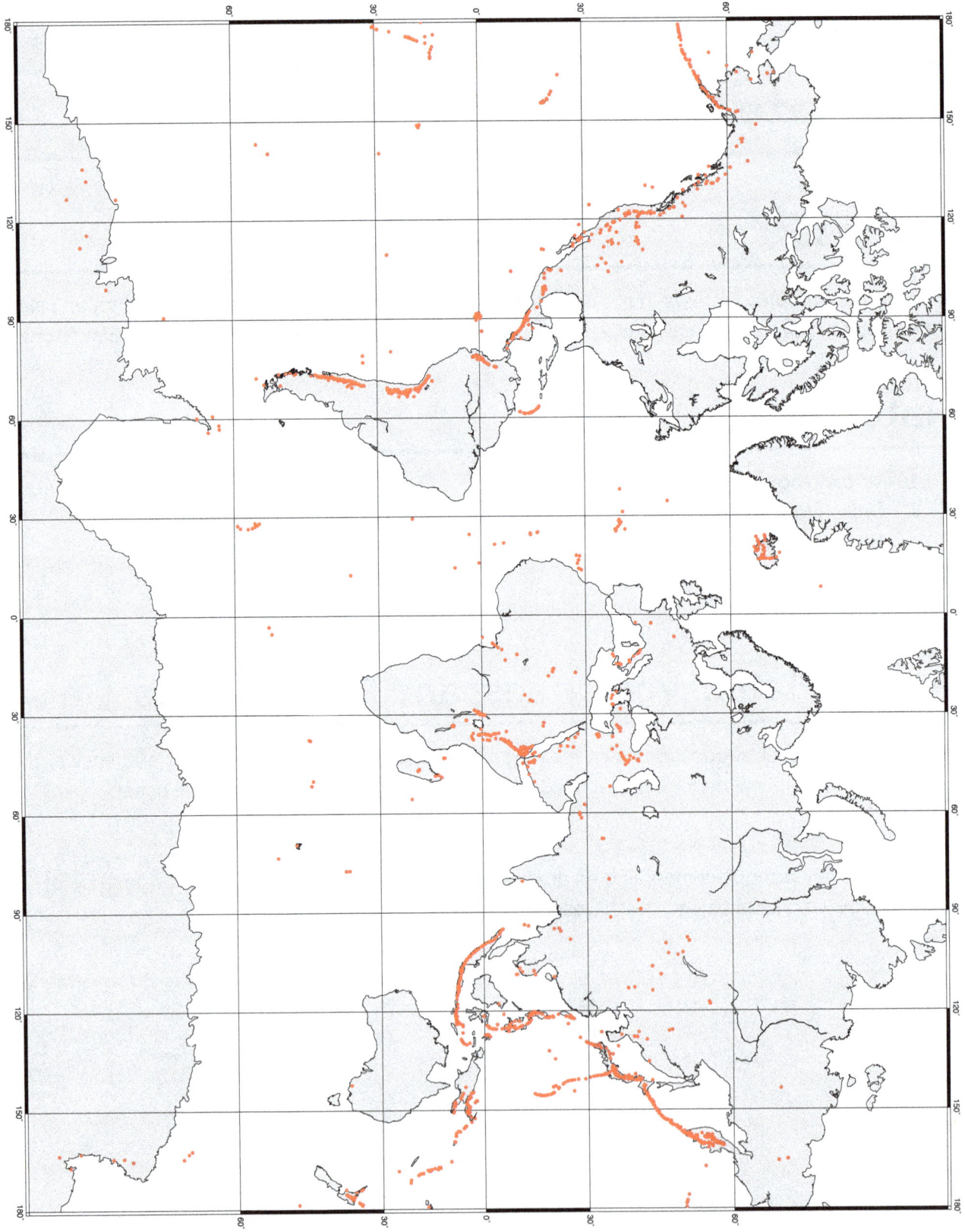

Figure 2.2

SCIENTIFIC SPECIALTY: SEISMOLOGY

Earthquake Locations 1990 - 1996 (Magnitudes 4 and greater)

Color indicates depth: Red 0-33 km, Orange 33-70 km, Green 70-300 km, Blue 300-700 km

This map is part of "Discovering Plate Boundaries," a classroom
exercise developed by Dale S. Sawyer at Rice University (dale@rice.edu).
Additional information about this exercise can be found at
http://terra.rice.edu/plateboundary .

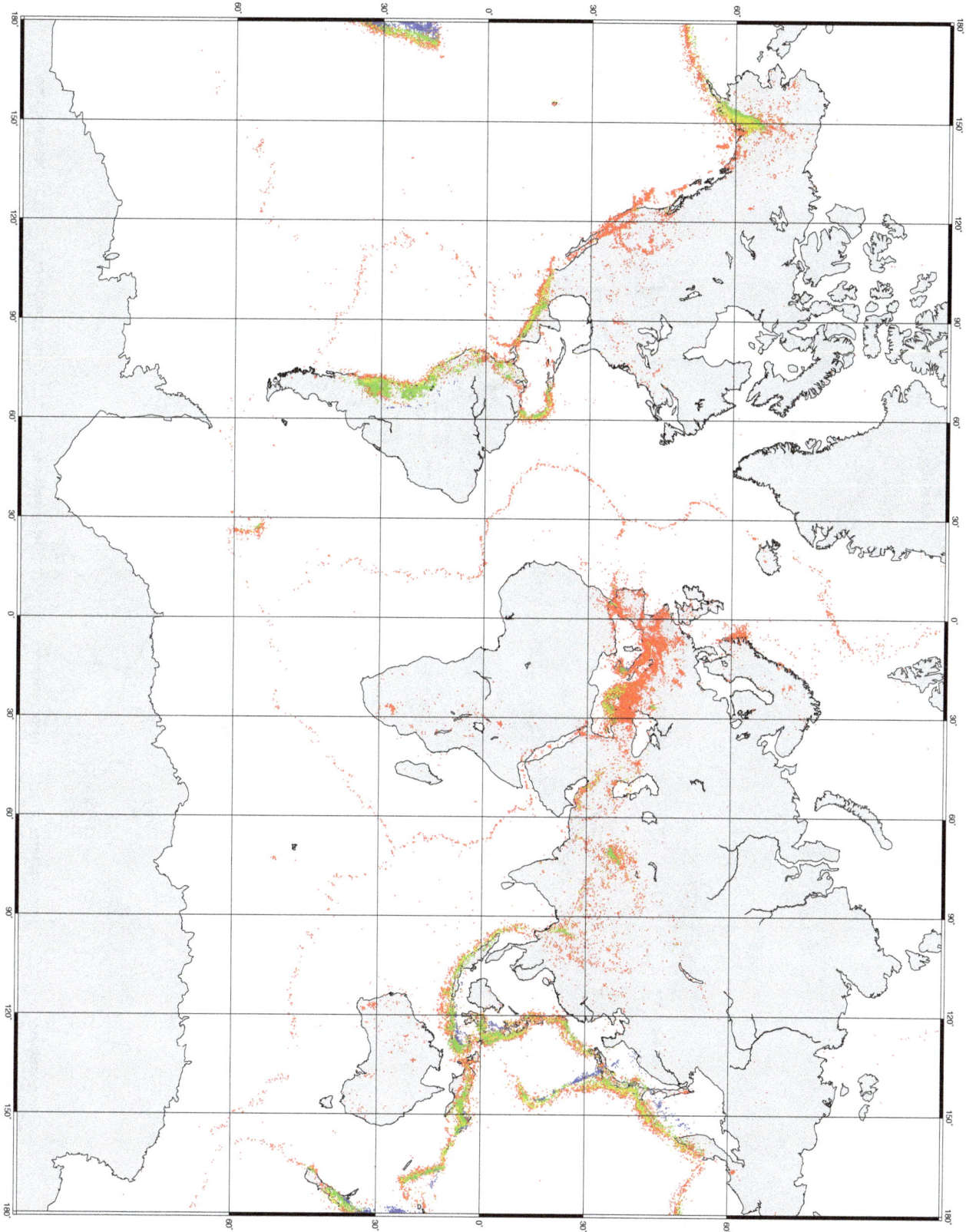

Figure 2.3

PLATE BOUNDARY MAP

This map is from Dietmar Mueller, Univ. of Sydney

This map is part of "Discovering Plate Boundaries," a classroom
exercise developed by Dale S. Sawyer at Rice University (dale@rice.edu).
Additional information about this exercise can be found at
http://terra.rice.edu/plateboundary .

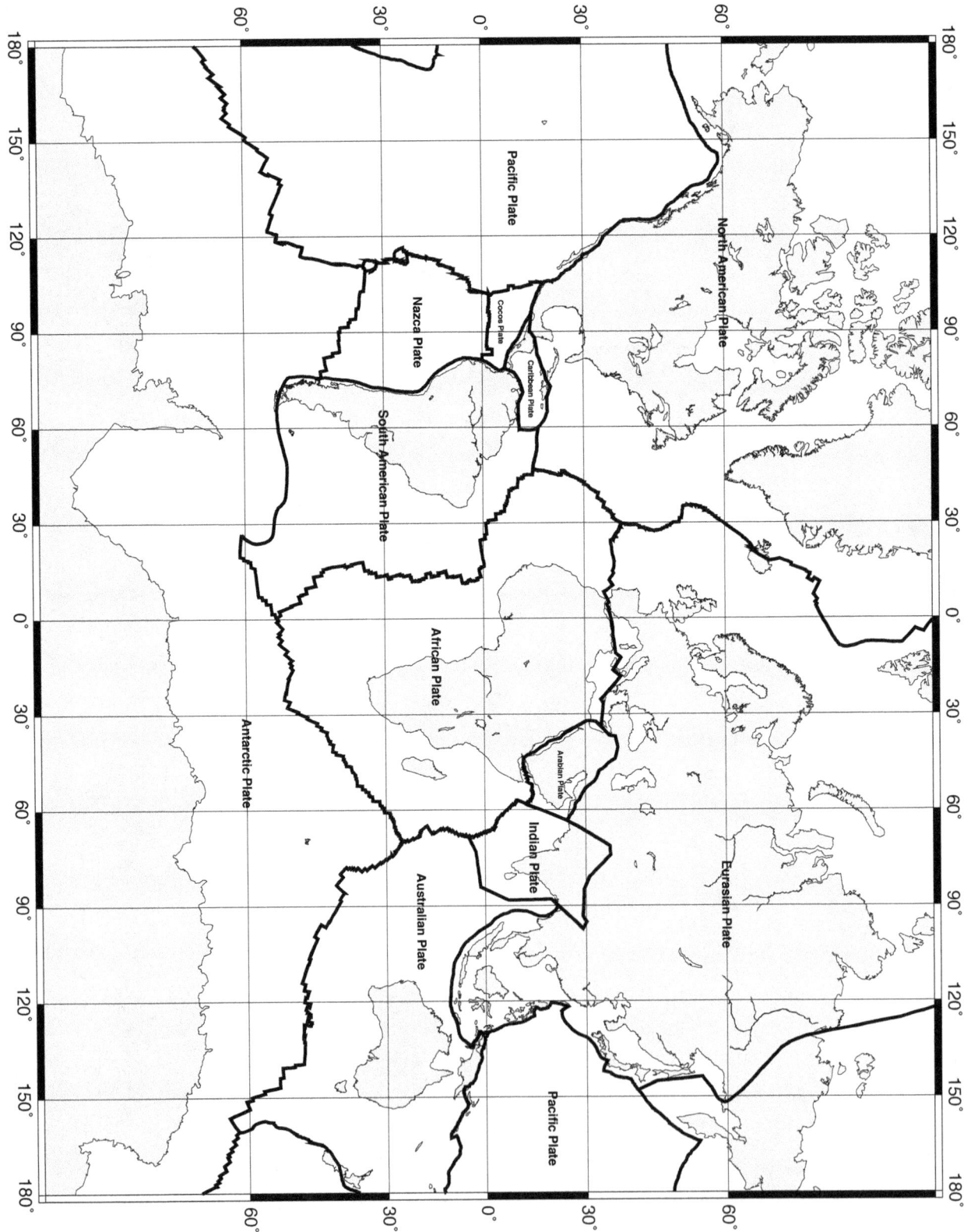

Q2. Based on the location of linear shallow earthquakes, describe the process and plate boundary most likely associated with these shallow earthquakes.

Q3. Using the earthquake map (figure 2.2), find an area in the oceans where there is an occurrence of deeper earthquakes (blue and green dots). Describe the location of the area.

Q4. Based on the location of deep earthquakes, describe the process and plate boundary most likely associated with these deep earthquakes.

Q5. Using the volcano map (figure 2.1), look at the western Pacific Ocean around and south of Japan. Based on the location of these volcanoes, describe the process and plate boundary most likely associated with these volcanoes.

Q6. Using the volcano map, find a singular volcano in the ocean that is separated from any other volcano. Describe this volcano's location.

Q7. How do you think this singular volcano is formed? Why are there no other active volcanoes near it?

Q8. On the map provided and using the information from both volcano (figure 2.1) and earthquake (figure 2.2) maps, draw the following features:

- ❑ Mid-Atlantic Ridge
- ❑ Java Trench
- ❑ Hawaiian Hot Spot
- ❑ East Pacific Rise (rise and ridge mean the same thing)
- ❑ Puerto Rico Trench
- ❑ Iceland Hot Spot

Q9. Using the information from above, why is Iceland the only large island in the Atlantic?

Part 2: Hot Spot Chains

Hot spot activity can be used to find out how fast a plate is moving. Using data from the Hawaiian Hot Spot, you will be calculating how fast the Pacific Plate is moving and its direction of movement.

Q10. Plot the data from table 2.1 on the provided graph paper (p. 16). In this graph, you will be plotting age on the x-axis and distance on the y-axis. After plotting these points, make sure to label your axes and give your graph a name.

Table 2.1

Volcano #	Volcano name	Age (Ma)	Distance (km)
1	Kilauea	0.20	0
2	Mauna Kea	0.38	54
3	Kohala	0.43	100
4	East Maui	0.75	182
5	Kahoolawe	1.03	185
6	West Maui	1.32	221
7	Lanai	1.28	226
8	East Molokai	1.76	256
9	West Molokai	1.90	280
10	Koolau	2.60	339
11	Waianae	3.70	374
12	Kauai	5.10	519
13	Niihau	4.89	565
14	Nihoa	7.20	780
15	unnamed 1	9.60	913

Q11. Draw a best fit line of your data (if you don't know what this is, ask your instructor). Once you have drawn this line, find the slope of your line (rise over run). The slope of this line is your plate motion rate in kilometers per year (km/yr should be in millions, so make sure you have a large number on the bottom with a lot of zeros). Write the slope below.

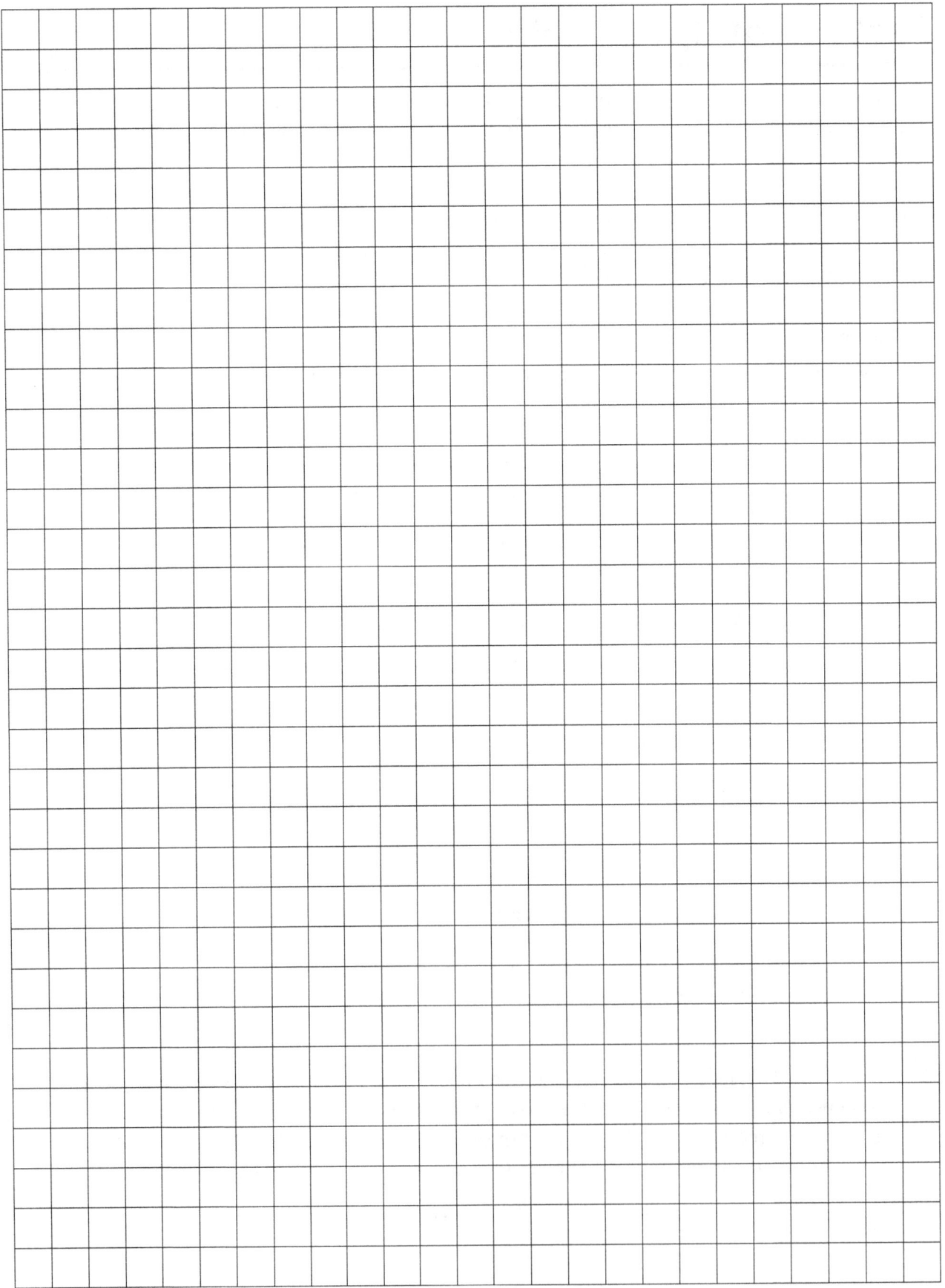

Q12. Using your rate from Q11, convert the number of kilometers to centimeters and calculate the rate of plate motion in centimeters/year (cm/yr). *Note: 100cm = 1m and 1,000m = 1km.*

Q13. The average rate of motion for the Pacific Plate is 9 cm/yr. How close was your calculated rate of plate motion? If it was not close, why do you think your answer was different?

Part 3: Calculating Seafloor Spreading Rate

Seafloor spreading rate is calculated a little differently than average plate motion. Instead of a straight-forward motion, you would calculate what is called the half spreading rate and multiply that by 2. To do this you need to first convert your distance to centimeters. Once you have centimeters calculated, you will use the equation distance/time = rate (cm/yr). You will then multiply the rate by 2 to get the Full Seafloor Spreading Rate. (*Hint: 100cm = 1m and 1000m = 1km*)

Q14. The oldest portion of the Juan De Fuca plate off the US west coast is from the crest of the Juan de Fuca Ridge and is 8 million years old. Port Orford, Oregon, is 375 km from the ridge. Calculate the Full Seafloor Spreading Rate.

Q15. Another portion of the Juan de Fuca plate is made up of the Gorda Ridge in Northern California. The oldest crust from this ridge is 6 million years old, and Cape Mendocino is 275 km from the ridge. Calculate the Full Seafloor Spreading Rate.

Q16. Was there a difference between your rates in questions 1 and 2? Should there be different rates if the Juan de Fuca plate is being created by two separate mid-ocean ridges?

Part 4: Comparing Oceans and their Features

Using the provided ocean floor topographic map, answer the following questions:

Q17. There are two separate subduction zones in the Atlantic Ocean. What are they named?

Q18. What type of landform is very prominent in the center of the Atlantic Ocean? What is the name of this landform?

Q19. What is the name of the long feature that can be seen in the eastern Pacific Ocean, far off the coast of South America? What plate boundary is it associated with?

Q20. What is the long, oceanic feature that can be seen on the east coast of Japan? What type of plate boundary creates this landform?

Q21. What sort of process do you think created the islands found in the western Pacific? What evidence do you have to determine the process?

Q22. Look for the Line Islands in the western Pacific. What sort of process created these islands that are not associated with a plate boundary?

Part 5: Density and Plate Tectonics

For this activity, you will be measuring the density of rock associated with the two types of Earth's crust.

Materials

- ❑ Rock sample of basalt
- ❑ Rock sample of granite
- ❑ Graduated cylinder
- ❑ Scale

Procedure

1. Weigh each sample on the balance and record the mass in the table below.
2. Fill the cylinder with water and record the initial water level here:
3. Gently place the rock sample in the cylinder. Avoid splashing. Record the new level of water here:

The volume of the rock sample is equal to the water it displaced, so find the displacement by subtracting the initial water level from the ending level of water. Record the volume of water for each sample in the graduated cylinder in the table below. *Note: 1 mL = 1 cm³.*

4. Fill out table 2.2:

Table 2.2

Rock Type	Mass (g)	Volume (cm3)	Density (g/cm3)
Basalt			
Granite			

Q23. Which sample composes the oceanic crust? Which sample composes the continental crust?

Q24. The true density of basalt is 3.0 g/cm^3 and the true density of granite is 2.7 g/cm^3. How close were your measurements? What could account for any error you see?

Q25. Earth's overall density is 5.52 g/cm^3. How does Earth's density compare to the density of the rocks you measured? What does this tell you about Earth's internal structure?

3 Marine Sediments

Purpose

To identify what type of sediments are found in ocean systems and the ways these sediments can be identified in terms of size, composition, and origin.

Materials

- ❑ 5 samples of sediment with different grain sizes
- ❑ Ruler
- ❑ Scale
- ❑ Slides of siliceous and calcareous organisms

- ❑ Microscope
- ❑ Weigh Boats
- ❑ Colored Pencils
- ❑ Calculator
- ❑ 3 sets of mixed sediments
- ❑ Set of Sediment Sieves

Part 1: Sediment Size

Using the sediment samples in figure 3.1 and rulers provided to answer the following questions.

Figure 3.1

Millimeters (mm)	Micrometers (µm)	Phi (Φ)	Wentworth size class	
4096		-12.0	Boulder	
256		-8.0		Gravel
			Cobble	Gravel
64		-6.0		Gravel
			Pebble	Gravel
4		-2.0		Gravel
			Granule	Gravel
2.00		-1.0		Gravel
1.00		0.0	Very course sand	Sand
			Course sand	Sand
1/2 0.50	500	1.0		Sand
			Medium sand	Sand
1/4 0.25	250	2.0		Sand
			Fine sand	Sand
1/8 0.125	125	3.0		Sand
			Very find sand	Sand
1/16 0.0625	63	4.0		Sand
			Course silt	Silt
1/32 0.031	31	5.0		Silt
			Medium silt	Silt
1/64 0.0156	15.6	6.0		Silt
			Fine silt	Silt
1/128 0.0078	7.8	7.0		Silt
			Very fine silt	Silt
1/256 0.0039	3.9	8.0		Silt
			Clay	Mud
0.0006	0.06	14.0	Clay	Mud

Q1. Identify the Wentworth size class for each sample based on the Wentworth Scale using a ruler and by touch:

    a. _____

    b. _____

    c. _____

    d. _____

    e. _____

Q2. Think about the energy necessary to move the above sediments. Based on grain size, where in the marine system do you think gravel can be found? Sand? Silt and Clay? (use the locations and depth including coastal stream, beach, and deeper ocean)

Q3. Why are sediments sorted in the way your describes in the previous question?

Part 2: Sediment Sorting

For this portion, three lab groups will use the different sediment shakers to obtain the necessary data to fill out the following table. **Do not measure the sediment sizes for all samples.** If your group measures the sediment sizes of a sample, go to the front of the lab area and write your findings on the board.

Once all sediments have been measured, there is no need to reuse the sediment shakers. To find the weight percent of the specific sediment size versus the entire sample (wt%), divide the weight of the portion of the sample by the total weight of the sample and multiply by 100 (wt/starting wt) * 100).

Materials
- ❑ 3 sets of mixed sediments
- ❑ Scale
- ❑ Weigh boats
- ❑ Calculator

Table 3.1

Sieve	Sample 1	Sample 2	Sample 3
	Starting Weight:	Starting Weight:	Starting Weight:
Coarsest (3.4mm)	_____wt _____wt%	_____wt _____wt%	_____wt _____wt%
Second Coarsest (.86mm)	_____wt _____wt%	_____wt _____wt%	_____wt _____wt%
Third Coarsest (.38mm)	_____wt _____wt%	_____wt _____wt%	_____wt _____wt%
Fourth Coarsest (.15mm)	_____wt _____wt%	_____wt _____wt%	_____wt _____wt%
Bottom Pan	_____wt _____wt%	_____wt _____wt%	_____wt _____wt%
Total of all weights			

Q4. Did your total of all weights equal your starting weights? If not, why do you think they are different?

Q5. Create a bar graph of your results using the charts below:

Sample #1

90	
80	
70	
60	
50	
40	
30	
20	
10	
0	

Coarsest Sieve   Second Sieve   Third Sieve   Fourth Sieve   Pan

Sample #2

90	
80	
70	
60	
50	
40	
30	
20	
10	
0	

Coarsest Sieve   Second Sieve   Third Sieve   Fourth Sieve   Pan

Sample #3

90				
80				
70				
60				
50				
40				
30				
20				
10				
0				

Coarsest Sieve   Second Sieve   Third Sieve   Fourth Sieve   Pan

Q6. Which of the samples was the best sorted (i.e. has the fewest sediment sizes represented)?

Q7. Which of the samples was the worst sorted (i.e. has the most sediment sizes represented)?

Q8. Think about how sediments are deposited in the marine environment. Based on sorting, which sample do you think you would find in a near shore river system? On a beach? In the deeper ocean?

Part 3: Oozes

Use the microscopes to view the plankton slides.

Q9. Sketch and describe both plankton slides in table 3.2.

Table 3.2

Draw the sample	A.	B.
Name		
Composition		
Relative Depth		

Q10. What is the origin of this sediment? Use the origin classification name.

Q11. What is the Carbon Compensation Depth (CCD)? Why can't calcareous oozes form below this depth?

Part 4: Sediment Core Analysis

Use table 3.3 and figure 3.1 to graph the core samples on the following graph where the x-axis represents %CaCO$_3$ (calcium carbonate) and the y-axis represents water depth. Use two different colors to plot the sediment's cores, one color for the Atlantic Ocean and one color for the Pacific Ocean. Make sure to create a key.

Table 3.3

Core #	% CaCO$_3$ in upper portion	Water depth (m)
1	95%	1900
2	85%	2000
3	13%	5200
4	65%	4200
5	8%	5800
6	1%	9000
7	3%	8000
8	10%	5500
9	75%	3600
10	90%	1000

Figure 3.1

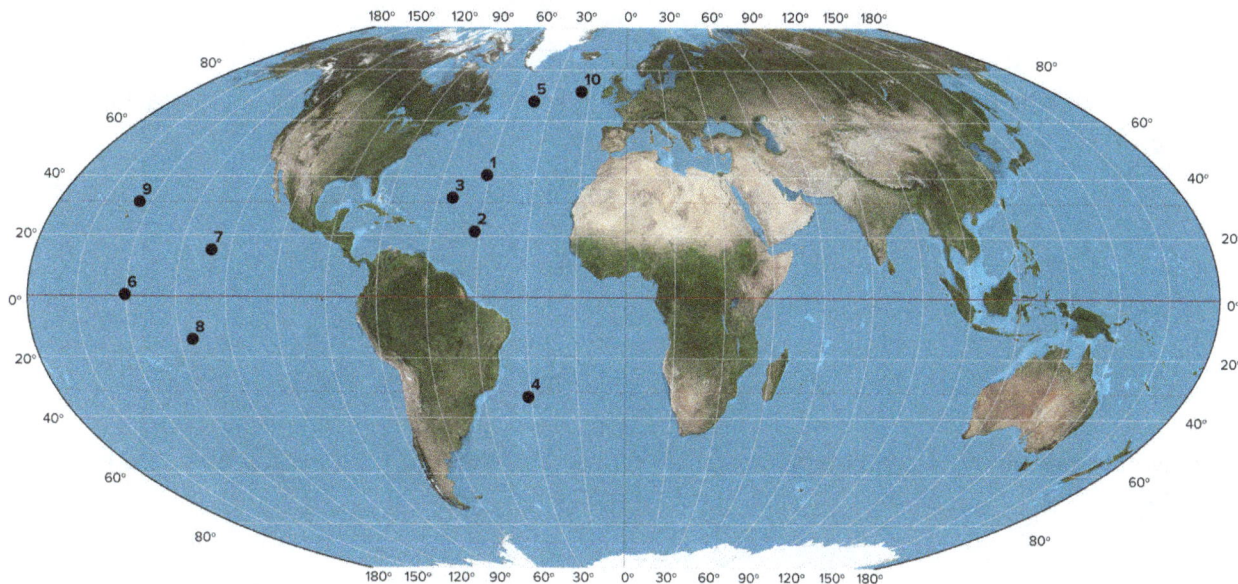

Q12. Plot the data from the cores on the graph in figure 3.2:

Figure 3.2. % CaCO₃ in top 10 cm

Depth (m)	10%	20%	30%	40%	50%	60%	70%	80%	90%	100%
1000										
2000										
3000										
4000										
5000										
6000										
7000										
8000										
9000										

Q13. What do you notice about the percentage of CaCO₃ with depth? How do you explain this pattern?

Q14. Which ocean shows a greater percentage of CaCO₃? What sort of feature could account for this difference?

Q15. If you were to take a core sample directly off the coast of Oregon, would you expect a high level or low level of CaCO₃? Explain your answer.

Q16. Notice the elevated percentage of CaCO₃ in core #9. Why do you think the percentage is high?

4 Water

Purpose

To understand the physical and chemical properties of ocean water.

Materials

- ❏ Bouncy ball
- ❏ Small plastic cylinder
- ❏ Wood block
- ❏ Small pyramid
- ❏ Plastic tub
- ❏ Salt
- ❏ Calculator
- ❏ Colored pencils
- ❏ Rectangular tank with

divider
- ❏ Colored saltwater
- ❏ Colored freshwater
- ❏ Colored cold water
- ❏ Colored warm water
- ❏ Flask
- ❏ One hole stopper
- ❏ Long glass tube
- ❏ Food coloring

- ❏ Warm water
- ❏ Breaker
- ❏ Ice cubes
- ❏ Beaker
- ❏ Hot plate
- ❏ Thermometer
- ❏ Thermometer clip
- ❏ Graph paper

Part 1: Density

For this experiment, you will hypothesize which of the following objects should sink and which items should float in water. To do this, calculate the volume of each material and measure the mass to find the density. Show your work.

Materials

- ❏ Bouncy ball
- ❏ Small plastic cylinder
- ❏ Wood block
- ❏ Small pyramid
- ❏ Plastic tub
- ❏ Calculator

Q1. Sphere: $V = c^3/(6\pi^2)$

    a.   Weight of bouncy ball (g): _____

    b.   Volume of bouncy ball (cm3): _____

    c.   Density of bouncy ball (g/cm3): _____

Q2. Cylinder: $V = \pi r^2 h$

    a.   Weight of plastic cylinder (g): _____

    b.   Volume of plastic cylinder (cm3): _____

    c.   Density of plastic cylinder (g/cm3): _____

Q3. Cube: $V = L \times W \times H$

    a.   Weight of wood cube (g): _____

    b.   Volume of wood cube (cm3): _____

    c.   Density of wood cube (g/cm3): _____

Q4. Pyramid: V = (L × W × H)/3

    a.   Weight of rubber pyramid (g): _____

    b.   Volume of rubber pyramid (cm3): _____

    c.   Density of rubber pyramid (g/cm3): _____

Q5. Now that you have found the density of each object, hypothesize which ones you think will float and which you think will sink. Why do you think that?

Q6. Now, using the plastic tub, test your hypothesis. Were you correct? If no, why not?

Q7. What do you notice about the density of those that float and those that sink? How does this allow you to estimate the density of water?

Q8. If you were to place these objects in very salty water, hypothesize whether you would see the same results. Explain your reasoning.

Part 2: Water Density

Materials

- ❏ Rectangular tank with divider
- ❏ Colored saltwater
- ❏ Colored freshwater
- ❏ Colored cold water
- ❏ Colored warm water

Q9. Make sure the divider is placed into the tank correctly. For the first trial, place saltwater in one side of the tank and freshwater in the other side. What do you think will happen when you remove the divider? Write your hypothesis here.

Q10. Now remove the divider. Was your hypothesis correct?

Q11. Pour out the water from your first trial and replace the divider correctly. For the second trial, place cold water in one side of the tank and warm water in the other side. What do you think will happen when you remove the divider? Write your hypothesis.

Q12. Now remove the divider. Was your hypothesis correct?

Q13. Describe the relationship between temperature and density.

Figure 4.1

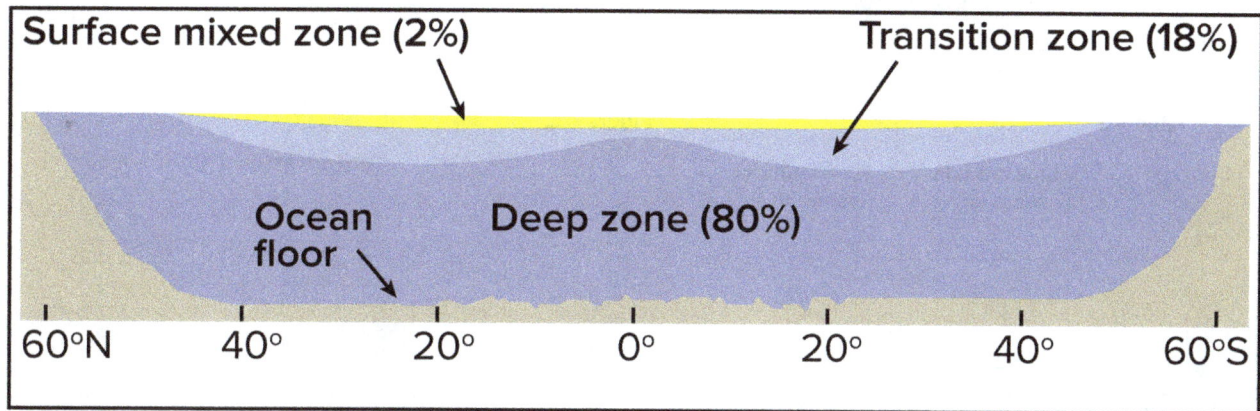

Surface mixed zone (2%)   Transition zone (18%)

Ocean floor   Deep zone (80%)

60°N   40°   20°   0°   20°   40°   60°S

Q14. Complete table 4.1 in terms of general density and temperature relationship based on the layers shown in figure 4.1. Which layer has the lowest, middle, and highest temperature and density?

Table 4.1

Ocean Zone	Temperature	Density
Surface Zone (Mixed Layer)		
Transition Zone		
Deep Zone		

Q15. Use table 4.2 to plot the pycnocline of high and low latitude water density on figure 4.2. Use different colors for high and low latitudes. Label the surface zone, pycnocline, and deep zone.

Table 4.2

Depth	Low latitude density (g/cm³)	High latitude density (g/cm³)
0	1.0250	1.0280
250	1.0260	1.0280
500	1.0270	1.0280
750	1.0275	1.0280
1000	1.0280	1.0280
2000	1.0280	1.0280
3000	1.0280	1.0280
4000	1.0280	1.0280

Figure 4.2

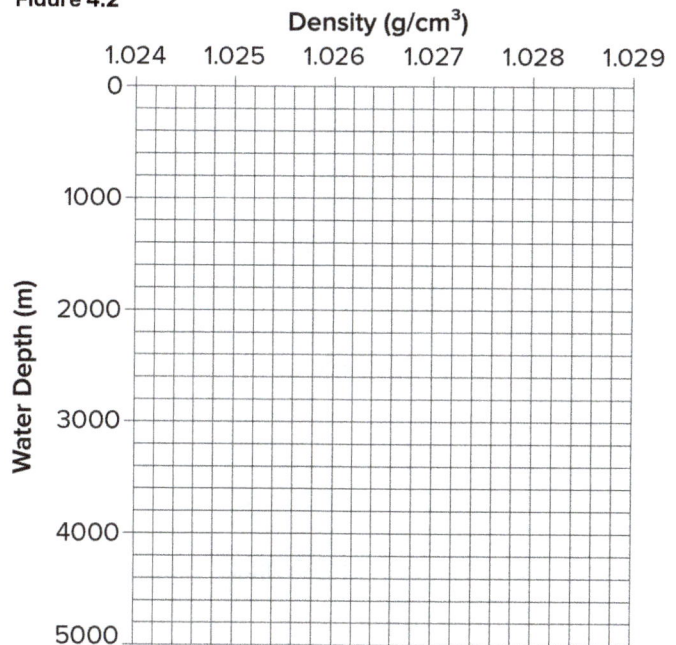

Density (g/cm³)

1.024   1.025   1.026   1.027   1.028   1.029

Water Depth (m)

0
1000
2000
3000
4000
5000

Q16. Use table 4.3 to plot the thermoocline of high and low latitude water temperature in figure 4.3. Use different colors for high and low latitudes. Label the surface zone, thermocline, and deep zone.

Table 4.3

Depth	Low latitude temp (oC)	High latitude temp (oC)
0	25	2
250	24	2
500	22	2
750	13	2
1000	4.5	2
2000	2	2
3000	2	2
4000	2	2

Figure 4.3

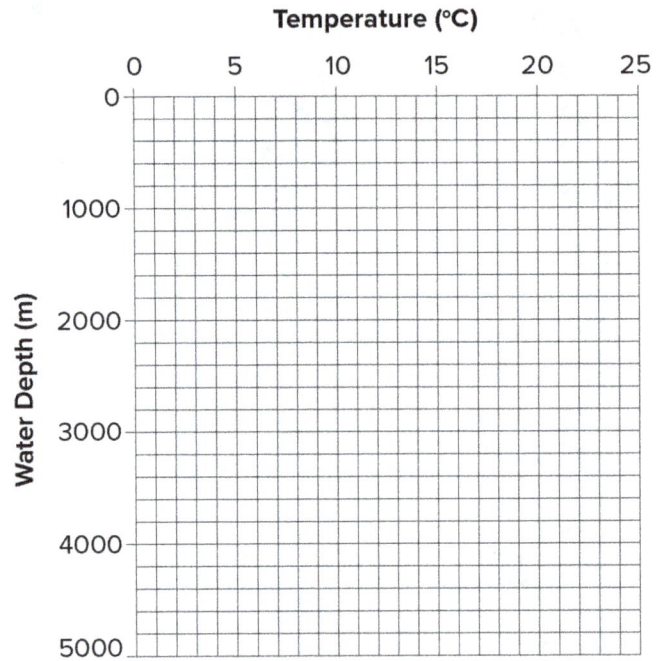

Q17. Why are there no thermocline nor pycnocline in higher latitudes?

Part 3: Warm Water

Materials
- [ ] Flask
- [ ] One hole stopper
- [ ] Long glass tube
- [ ] Food coloring
- [ ] Warm water
- [ ] Breaker

Q18. Fill the flask with water and add some food coloring. Once you've done that, mix the water up so the coloring spreads throughout the flask. Then push the glass tube through the single-hole stopper until it sits right at the surface of the water (not in the water). Write a hypothesis below about what you think will happen as the water in the flask is heated. Explain your hypothesis.

Q19. Place your apparatus on the hot plate and turn it onto a lower heat. Once your water is steaming (not boiling), observe the lab setup for 5 minutes and write your observations below.

Q20. Based on your observations, what should happen to ocean levels as the world warms?

Q21. What other processes can you think of that would cause ocean levels to rise? What can cause ocean levels to fall? Think of two examples for each question.

Part 4: Latent Heat of Fusion and Vaporization

For this experiment, you will watch some phase changes and record temperature data to create a graph of the temperature changes by following the listed steps below.

Materials

- Ice cubes
- Beaker
- Hot plate

- Thermometer
- Thermometer clip

Procedure

1. Fill your beaker about half full of ice.
2. Place the thermometer into the beaker by sliding it through the thermometer clip and attaching it to the side of the beaker with ice. The thermometer should not touch the bottom or side of the beaker. You may have to gently hold it in place.
3. After about one minute in the ice, record the temperature in table 4.4.
4. Place the beaker with ice cubes on the hot plate and turn the dial between midway and high.
5. After every minute, record the new temperature in table 4.4 for phase change. Add a note when the ice is fully melted and another note when steam begins to first appear.
6. Continue to observe the beaker until you have recorded enough temperature readings to fill table 4.4.
7. Present your data on the provided graph paper (p. 38) as a line graph. Label the x-axis "time" and the y-axis "temperature." Clearly label where the ice fully melted, where steam began forming, and any other important information recorded with your data.

Q22. How did temperature change after all the ice melted?

Q23. How did the temperature increase after the water began steaming?

Q24. Explain why the temperature did not change in a linear fashion? What properties of water can explain the plateaus recorded as the state of water changes?

Table 4.4

Time (min)	Temperature (°C)	Other Observations

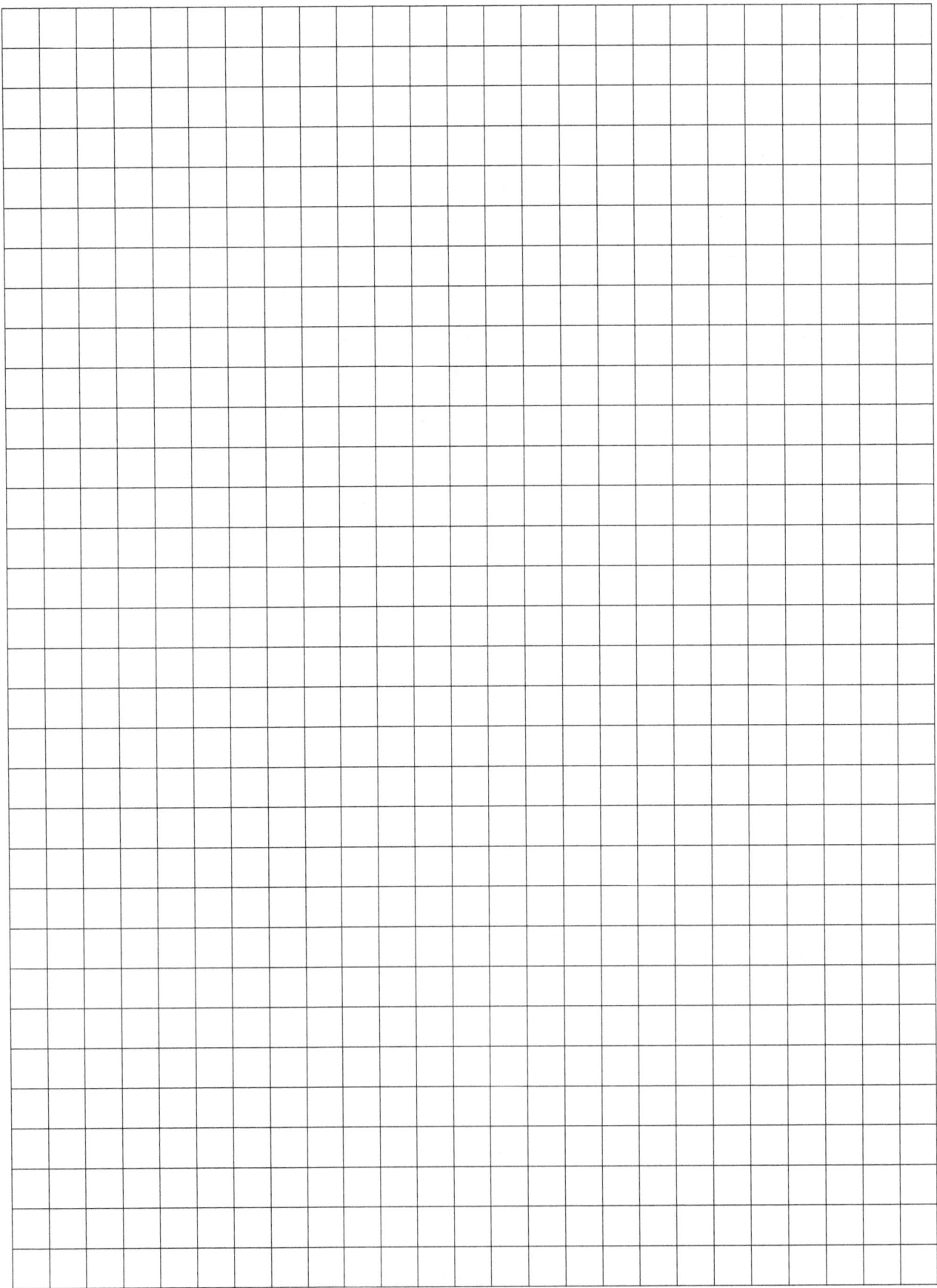

| 5 | **Ocean Currents** |

Purpose

To understand how surface and deep ocean currents form and flow.

Materials

- ❑ Clear plastic bin
- ❑ Cold colored water
- ❑ Turkey baster
- ❑ Fan
- ❑ Laptop or Chromebook

- ❑ Paper plate
- ❑ Pencil
- ❑ Pipette
- ❑ Colored water

Part 1: Nike Shoe Spill

During a large Pacific storm on May 27, 1990, a freighter called the Hansa Carrier was sailing from South Korea to the United States when a wave swept 21 large shipping containers off the deck. Four or five of these containers opened and released 60,000 Nike shoes into the North Pacific Ocean. These shoes washed ashore in various places, and we can use their landing locations to understand circulation patterns in the Pacific Ocean.

1. Use table 5.1 to plot the locations of where shoes were found on the map (figure 5.1).

Table 5.1

Incidents / Number of Shoes	Date	Location
Shoe Spill	May 27, 1990	48°N, 161°W
200 shoes recovered	November 1990	49°N, 126°W
100 shoes recovered	February 1991	53°N, 131°W
200 shoes recovered	February 1991	47°N, 125°W
250 shoes recovered	March 26, 1991	59°N, 139°W
150 shoes recovered	April 4, 1991	44°N, 124°W
200 shoes recovered	May 9–10, 1991	40°N, 124°W
200 shoes recovered	May 18, 1991	55°N, 130°W
Several recovered	January, 1993	19°N, 155.5°W
Several recovered	January 1994	32°N, 132°E
Several recovered	April 1996	54°N, 133°W

Figure 5.1

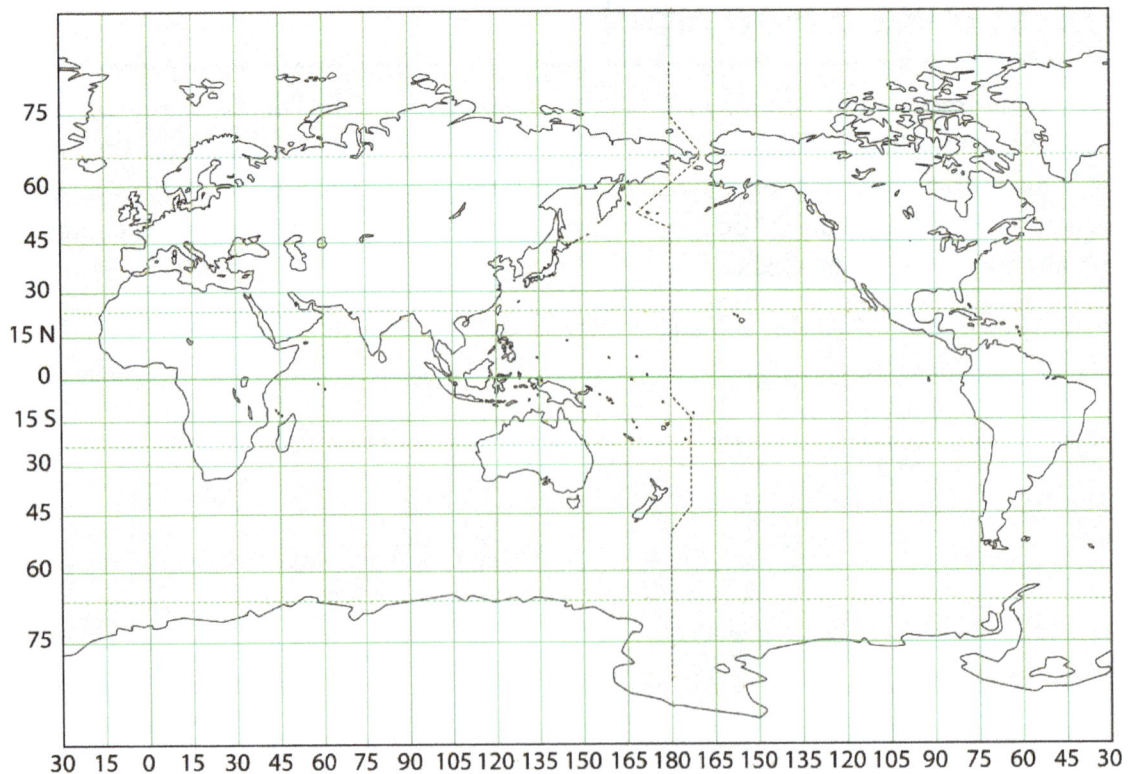

Q1. Add arrows between the dots on figure 5.1 to show how the shoes were moving. Describe in general how the shoes moved around the Pacific Ocean.

Q2. Using a map showing the world's ocean currents (provided by your instructor or found in your textbook), label the various ocean currents that carried the shoes around the Pacific Ocean on figure 5.1.

Q3. The distance between the spill and the first deposit of shoes is about 1,500 miles. Calculate the speed at which the shoes travelled in miles per day and miles per hour (use November 1 to estimate the number of days). Show your work.

Q4. The distance from California to Hawaii is approximately 2,400 miles. Calculate the how fast the shoes drifted between these two locations in miles per day and milers per hour (use the data point closest to the coast of California and the islands of Hawaii to estimate the number of days). Show your work.

Q5. What are the names of the two calculated currents from question 3 and 4. Which is the fastest?

Part 2: Upwelling

For this activity, you will recreate upwelling conditions found in the ocean along coastlines.

Materials
- ❑ Clear plastic bin
- ❑ Cold colored water
- ❑ Turkey baster
- ❑ Fan

Procedure
1. Fill the plastic bin ¾ full with regular tap water.
2. Fill the turkey baster with the colored cold water.
3. Place the turkey baster on the bottom of the bin, trying not to release the colored water. Once the tip of the baster is placed close to the bottom of the bin, *slowly* drain the cold water on the bottom of the tank.
4. Wait a few minutes for the cold water to spread out along the bottom of the bin. Once it has diffused a bit turn on your fan and hold it directly against one of the short edges of the bin with the fan blowing over the surface of the water away from that edge of the bin.
5. Blow air across the surface of the bin and observe the water in the bin through the side, not from the surface of the bin.

Q6. Draw and label a diagram of what you observed during your experiment.

Q7. If the deeper lower ocean waters are colder and therefore denser, how is that cold waters can rise up to the surface?

Q8. Based on the motion of the California Current and Ekman Transport, is upwelling occurring on the Oregon coast? What other parameters could you test in our coastal waters to see if upwelling was occurring?

Part 3: Sea Surface Temperature

Visit this website or scan the QR code to answer the following questions:
https://iridl.ldeo.columbia.edu/maproom/Global/Ocean_Temp/Monthly_Temp.html

Q9. Describe where the warmest and coldest temperatures are for the current map.

Q10. Once you've recorded your initial observations, hover your cursor over the map and when the menu bar shows up, click on the date to change it to 2015 (don't erase and rewrite the year, just reduce the numbers using the cursor). Use the left-hand arrow to find January 2015, then scroll through images for the entire year. Discuss how the sea surface temperatures change over a whole year, paying close attention to the equator, and especially the coast of South America.

Q11. Hover over the map and when the menu bar shows up click on the date and change the date to 2016 like you did for the last question. Use the left-hand arrow to find January 2016 and then scroll through images for the entire year. Discuss how the sea surface temperatures change over a whole year paying close attention to the equator, and especially the coast of South America. How are these temperatures different from 2015?

Q12. 2015–2016 had a specific oceanic event associated with it where ocean waters warmed in the eastern Pacific Ocean. What was the name of this event? How could these changes effect the climates on the west coast of North and South America? (think about evaporation)

Q13. For the next questions, examine the North Pacific in May 2016 (you can zoom in using the cursor by making a box around the area you would like to examine):

    a.   Compare the water off the Southern California coast to the rest of the Pacific at the same latitude. The water off the coast of Southern California is:

        ❏ Similar   ❏ Colder   ❏ Warmer

    b.   Compare the water off the east coast of the largest island of Japan to the rest of the Pacific at the same latitude. The water off the east coast of Japan is:

        ❏ Similar   ❏ Colder   ❏ Warmer

Q14. Hover over the map and click on the magnifying glass at the top right of the map to zoom out. Now zoom in closely to the Pacific coast of Oregon.

    a.   Compare the water off the coast of the Oregon to the rest of the Pacific in the same area. The water off the coast the Oregon is:

❑ Similar   ❑ Colder   ❑ Warmer

Q15. Compare the East Coast of the US to the West Coast of the US over the course of a year.

    a.   Which coast has warmer waters?

❑ East Coat   ❑ West Coast

    b.   How do ocean currents explain this difference?

Part 4: Coriolis Effect

For this activity, you will use a two-dimensional model of the globe to see how a medium behaves when the Earth spins.

Materials

- ❑ Paper Plate
- ❑ Pencil
- ❑ Pipette
- ❑ Colored water

Q16. Take your plate and pierce your pencil right through the center of the plate and move the plate a couple inches down your pencil. The plate is the Earth in this activity, so what direction would you spin the plate to mimic the spin of the Earth?

❑ Clockwise   ❑ Counterclockwise

Q17. Now that you know how to spin the model, use the pipette to drop a couple drops of colored liquid near the pencil. Once you have done that, start spinning the plate in the direction of spin of the Earth fairly quickly. Describe below what happens to the liquid.

Q18. Now, do the same thing, but spin the model faster than before. What happens to the water?

Q19. Do the same thing as described in question 2, but spin the model much slower than before. What happens to the liquid?

Q20. The top of the globe spins faster at the equator because it's the largest circle of latitude and objects have a larger inertia. Based on the previous questions, what portion of the globe is more affected by the Coriolis Effect?

❑ North Pole   ❑ Equator

Q21. What happens to the liquid if you turn the plate the other way and place water near the pencil?

6 Waves

Purpose

To understand wave characteristics and dynamics including wind, waves, tides, and tsunamis.

Materials

- ❑ Ruler
- ❑ Calculator
- ❑ Graph paper
- ❑ Plastic Bin

- ❑ Water
- ❑ Stopwatch
- ❑ Ruler
- ❑ Calculator

Part 1: Waves

Use the figure 6.1 diagram to answer the questions below. For each of your measurements on the diagram, 1cm = 1m. Show all your work and units for credit.

Figure 6.1

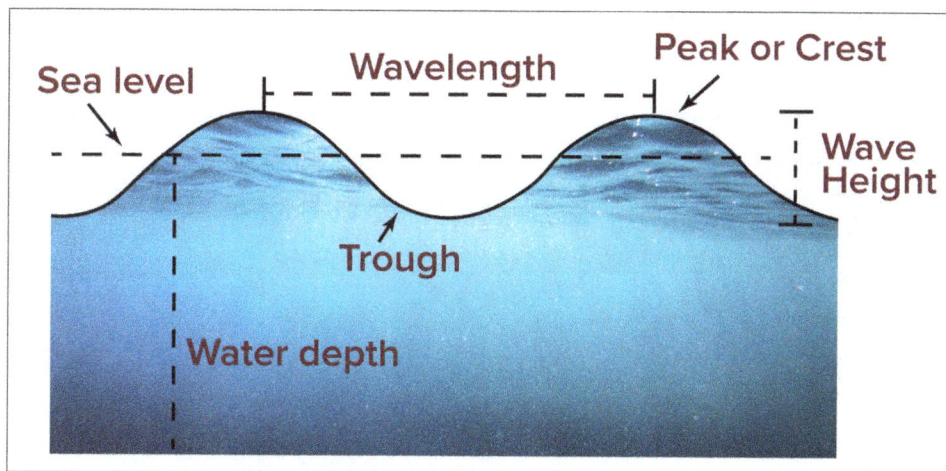

Q1. Measure the wave height and wavelength on this diagram, in meters.

Q2. Wave period is the time it takes for the wave to complete one cycle: e.g., one crest passes another or one trough passes another. What would the wave period be above if the wave above was moving at 0.6m/s? (*Hint: Try to cancel out meters.*)

Q3. Calculate the wave steepness (H/L). Would this theoretical wave break? *Note: Shallow-water waves usually begin to break when the ratio of wave height to wavelength is at or above a 1 to 7 ratio (H/L ≥ 1/7).*

Q4. Imagine this wave is stable enough to approach shore. What would happen to the wave height and wavelength at very shallow water depths? Be specific.

Part 2: Wave Depth and Speed

For this portion of the lab you will create waves in a plastic bin and observe their change in speed with depth for a shallow water wave.

Materials
☐ Plastic bin for water tank
☐ Water
☐ Stopwatch
☐ Ruler
☐ Calculator

Procedure

Read all the steps for this activity before carrying them out.

1. Measure the inside length (L) of the wave tank in centimeters and multiply it by 2. Record that number on table 6.1 because we are measuring the wave traveling across the tank and back (that's why L × 2).
2. Using a small ruler, fill the tank to a level of exactly 0.5 cm.
3. Create a wave by lifting the tank at one end into the air, letting the water settle on the other side of the tank. Then, set the tank down flat on the table, making sure it doesn't splash.
4. Use the stopwatch to measure the time it takes for a wave to travel the entire length of the bin twice (meaning back then forth). *Hint: Start your stopwatch after the wave has traveled to one end of the tank, or after the first wave.*
5. Repeat 4 more times and record each trial in table 6.1. Then find the average of the trials.
6. Repeat the steps above for a depth of 1 cm, 1.5 cm, 2 cm, and 2.5 cm. Record your data in table 6.1.
7. From the time averages, calculate the speed in centimeters per second (cm/sec) and record in table 6.1. (*Remember: speed = distance/time; where distance = L × 2*)

Table 6.1. Measured Speeds (don't forget to divide distance by time)

Times	Depths				
	0.5 cm	1 cm	1.5 cm	2 cm	2.5 cm
1					
2					
3					
4					
5					
Average					

8. Now calculate the theoretical speed using the following equation ($c = \sqrt{g \times d}$) and place your answers in table 6.2 where:
   - ❑ c = wave speed (celerity)
   - ❑ g = acceleration due to gravity (980cm/s²)
   - ❑ d = depth

Table 6.2

Depth	Average measured wave speeds (cm/sec)	Depth	Theoretical (calculated) wave speeds (cm/sec)
0.5 cm		0.5 cm	
1 cm		1 cm	
1.5 cm		1.5 cm	
2 cm		2 cm	
2.5 cm		2.5 cm	

6

49

9. With all of your data gathered, plot your results on the graph in figure 6.2 (this will be a line graph). Use green for your measured speed and orange for the provided theoretical speed.

Figure 6.2

60cm

40cm

20cm

0   0   0.5cm   1cm   1.5cm   2cm   2.5cm

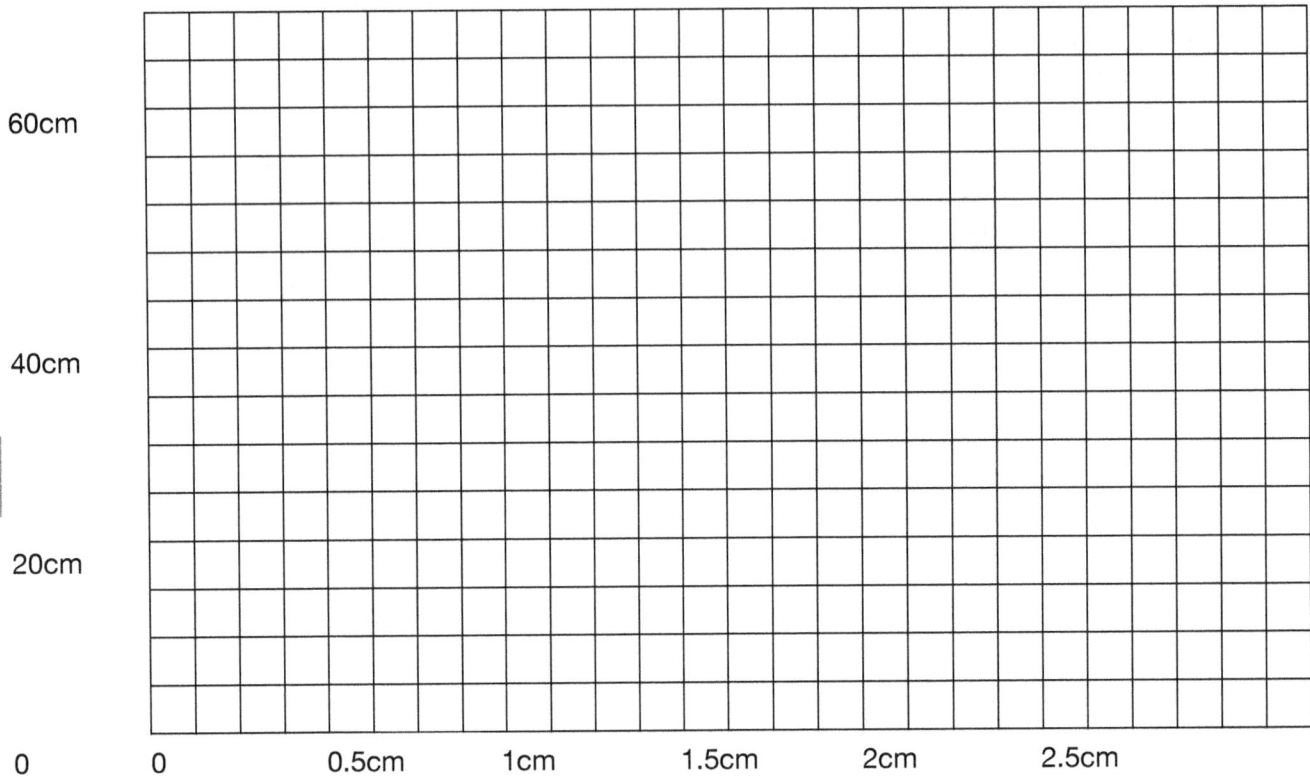

Q5. Based on your graph, how does speed relate to the depth of the water?

Q6. How accurate were your observed measurements compared to your calculated measurements?

Q7. Why do waves with a greater depth move faster than those with a shallower depth?

Part 3: Tides

For this portion of the lab, you will tell me which date has the highest tide, then graph the different tides for the month.

Table 6.3: Tide Data for South Beach Newport, Oregon (Meters above or below mean sea level)

Date	High Tide Level (m)	Low Tide Level (m)	Tidal Range (m)	Date	High Tide Level (m)	Low Tide Level (m)	Tidal Range (m)
January 1	0.54	0.49		January 21	0.93	−0.27	
January 2	0.59	0.46		January 22	0.96	−0.31	
January 3	0.66	0.31		January 23	0.98	−0.31	
January 4	0.75	0.13		January 24	0.99	−0.30	
January 5	0.85	−0.06		January 25	0.98	−0.27	
January 6	0.95	−0.23		January 26	0.94	−0.22	
January 7	1.05	−0.37		January 27	0.89	−0.16	
January 8	1.13	−0.46		January 28	0.82	−0.09	
January 9	1.18	−0.50		January 29	0.73	−0.01	
January 10	1.19	−0.48		January 30	0.60	0.08	
January 11	1.15	−0.42		January 31	0.45	0.17	
January 12	1.05	−0.31		February 1	0.28	0.23	
January 13	0.90	−0.17		February 2	0.72	0.07	
January 14	0.72	−0.02		February 3	0.78	−0.12	
January 15	0.53	0.12		February 4	0.87	−0.28	
January 16	0.34	0.25		February 5	0.98	−0.41	
January 17	0.72	0.15		February 6	1.07	−0.49	
January 18	0.77	0.00		February 7	1.13	−0.51	
January 19	0.83	−0.13		February 8	1.13	−0.46	
January 20	0.88	−0.22		February 9	1.07	−0.37	

Q8. Look over the data above and tell me which date is the highest high tide and the lowest low tide. What is the name for these types of tides?

Q9. Now look for the lowest high tide and the highest low tide and record those dates. What is the name for this type of tide?

Q10. Which day had the highest tidal range? Is this data close or a match for your answer in question 1?

Q11. Now graph the daily high and low tide levels on the provided graph paper. Use one color for high tides and another for low tides. **Make tide level the y-axis and date of the month the x-axis**. Make sure that you take into account that there are negative numbers in your data set when making your y-axis. Also, make sure to label your axes, name your graph, and provide a key for your colors.

Q12. Label the spring tides and the neap tides on your graph. How much time separates each event?

Q13. Based on the pattern of the graph, predict when the next spring tide will occur and when the next neap tide will occur.

Part 4: Tsunamis

Use the data in table 6.4, recorded by DART buoys during the 2011 Tohoku Japanese tsunami, and the figure 6.2 map to answer the questions.

Table 6.4.

Station	Arrival Time	Elapsed Time	Station	Arrival Time	Elapsed Time
21413	0700	1 hour	46407	1430	8.50
21414	0915	3.25	46408	1015	4.25
21415	0845	2.75	46409	1215	6.25
21416	0800	2.00	46410	1300	7.00
21418	0630	0.50	46411	1500	9.00
21419	0700	1.00	51406	1900	13.00
32411	2230	16.50	51425	1230	6.50
32412	0030	18.50	52402	0930	3.50
43412	1900	13.00	52403	1030	4.50
43413	2030	14.50	52405	0930	3.50
46402	1045	4.75	52406	1200	6.00
46403	1130	5.50	54401	1700	11.00
46404	1430	8.50			

Arrival times extracted from raw DART data: http://www.ngdc.noaa.gov/hazard/dart/2011honshu_dart.html

Figure 6.2

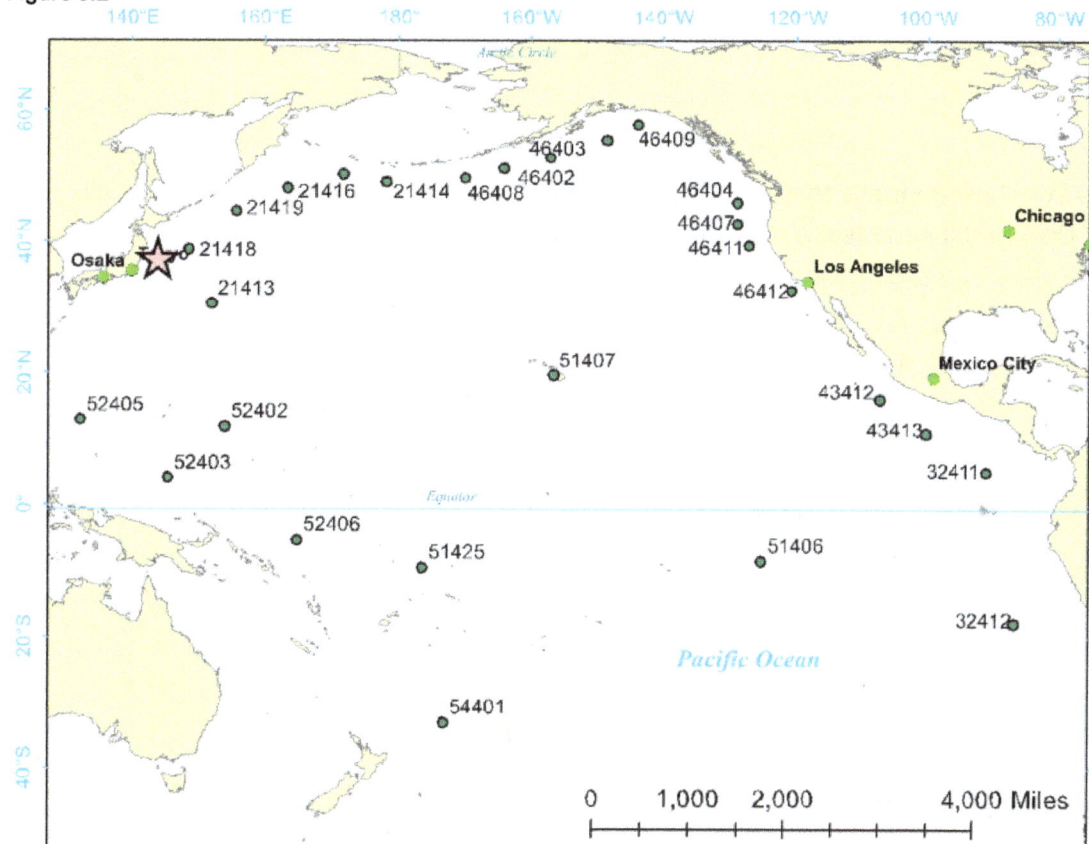

Q14. Use the data to draw the approximate location of the 2011 tsunami wave front after 5, 10, and 15 hours on the provided map. What approximate time did the wave hit Hawaii? Los Angeles?

Q15. Using the map scale to measure the distance from the epicenter and the time you estimated for the arrival of the wave in Los Angeles, how fast was the wave moving to be able to reach Los Angeles when it did? Report your answer in miles/hr.

Q16. Does your estimation above provide a good speed for the tsunami based on your knowledge of tsunamis? Why or why not?

Q17. What do you think happens to the strength of the tsunami as it moved across the Pacific Ocean? Do you think the tsunami could still be dangerous as it hit the Pacific Coast?

7   # Coastal Processes

Purpose

To identify coastal landforms and their creation processes using satellite photos, maps, and diagrams.

Materials

- ☐ Laptop/Chromebook
- ☐ Cape Sebastian quadrangle
- ☐ Colored pencils

Part 1: Coastal Landforms and Phenomena of Emergent Coast

Coastal processes are among the most dynamic geologic processes, where changes can happen in months if not years. Waves and storms are constantly eroding the beach and its related landforms. In this exercise, you will identify a number of common coastal landforms using Google Earth.

Using a computer, visit Google Earth and find the Oregon coast. Once you are there, zoom in a bit so you can start seeing the coastal features.

Q1. Look for the following coastal landforms and write the coordinates (lat/lon) where you found the landform and what city they are closest to (lat/lon can be seen at the lower right hand corner of the Google Earth screen):

    a.   Sea Stack: _____

    b.   Headland: _____

    c.   Sand Dunes:_____

    d.   Lagoon: _____

    e.   Sea Cliffs: _____

    f.   Bay: _____

    g.   Tombolo:_____

Q2. What is the main direction waves are travelling along the Oregon coast?

Q3. What is the longshore drift direction along the Oregon coast?

Q4. What evidence shows the direction of the longshore drift?

Q5. What do you think are the sources of sand along these beaches (think of where sand comes from inland of the beach)?

Q6. Where is the beach the widest in these images (give a direction or a description of a pattern you see)?

Q7. Is there evidence of human-induced coastal change along this coast? Briefly describe the evidence for human impacts on this coast.

Part 2: Cape Sebastian

Using the topographic map provided, answer the following questions to the best of your ability.

Q8. What is the ratio scale for this particular map?

Q9. Estimate the north-south and east-west dimension (true length) of this quadrangle in kilometers. Calculate its area in square kilometers (if you use the scale bar, make sure you use the right length).

Q10. Where are the beaches widest on this map? What would deposit so much sand in this location?

Q11. What is the contour interval on this map? What is the index contour interval (the elevation change between two dark likes)?

Q12. What is the elevation of Cape Sebastian? What type of landform is Cape Sebastian?

Q13. Name the major river located within your map's area. How is longshore drift affecting the mouth of this river?

Q14. The term "relief" is used to describe the difference between the highest and lowest elevations in an area. What is the total relief in the quadrangle?

Q15. What type of coastal feature is Cave Rock?

Q16. There are sand dunes mapped in Pistol River state park. Based on their orientation and what you know of sea breezes, what is the main direction of wind in this area?

Part 3: Responses to Erosion

Use the descriptions and pictures of the man-made erosion prevention features to answer the following questions.

Seawalls

Seawalls run parallel to the beach along the shore. They can be built with concrete, wood, steel, or boulders, and they are common on the Oregon coast. They essentially prevent the motion of sand from the shore. Seawalls usually displace the beach that they are built on and they prevent natural erosion from continuing inland. In fact, as waves hit the seawall, they reflect and drag sand from the beach out into the ocean. This can remove the beach entirely and prevent swash/backwash from occurring where seawalls are built.

Figure 7.1

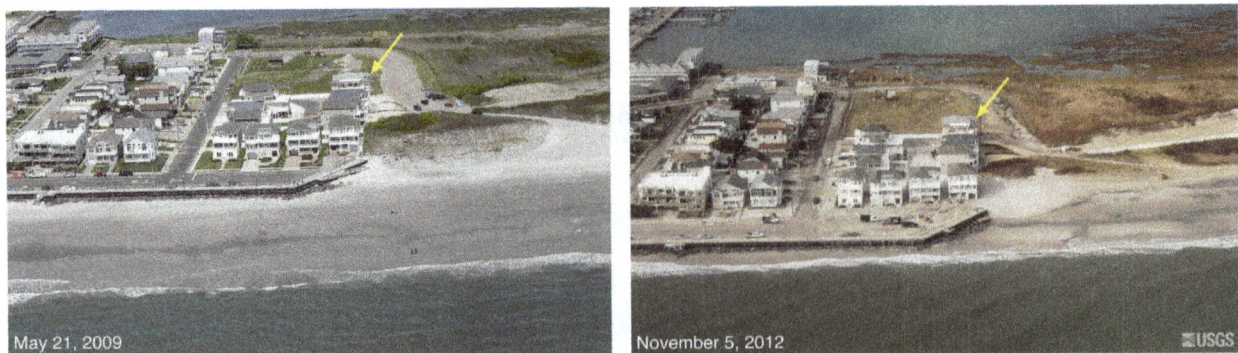

May 21, 2009   November 5, 2012   ≊USGS

Groins

A groin is a structure that is perpendicular to the beach. They are usually made of large boulders but can be made of concrete or steel. Groins are used to prevent the longshore drift from moving sand down the beach. Though it breaks up longshore drift, sand will accumulate on the up-drift side of the groin and be sand-starved on the down-drift side. Groins only slow down the longshore drift of sand, they do not prevent erosion.

Figure 7.2

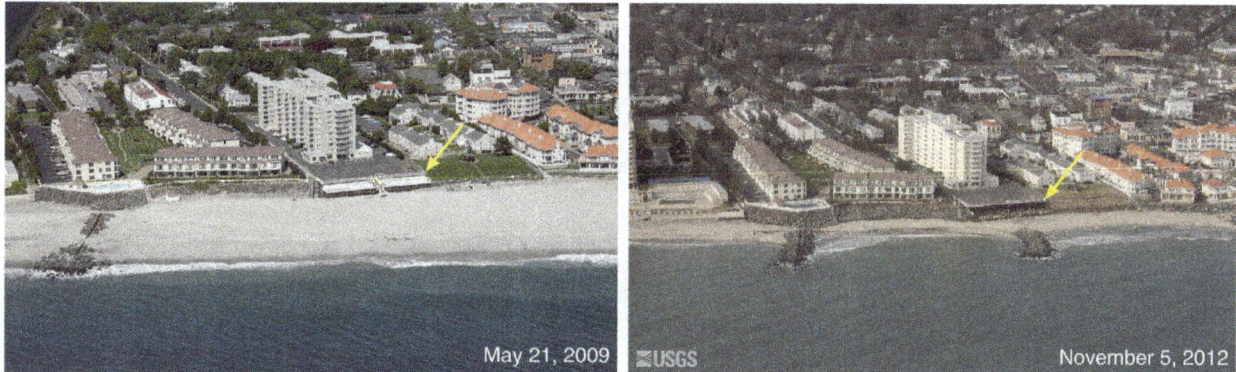

May 21, 2009 | USGS | November 5, 2012

Jetties

Jetties are built on either side of the mouth of a river flowing into the ocean. Jetties are made of boulders or concrete and are used to prevent the closing of the river channel by sand moved by longshore drift. Jetties completely redirect the longshore current and cause sand to accumulate on the up-drift side of the jetty, with the down-drift side becoming sediment-starved. Jetties are longer than groins and can permanently change the properties of a beach by affecting tidal circulation, wetlands, and other coastal features.

Figure 7.3

Breakwaters

A breakwater is built parallel to the shore and blocks waves from coming into the beach. Breakwaters slow waters in marinas and harbors. Breakwaters are made of boulders or concrete and can be submerged under water, elevated above water, or placed far off shore. The longshore current is interrupted by breakwaters and will change the characteristics of the beach. Sand will eventually accumulate towards a breakwater and downdrift sand will erode.

Figure 7.4

Q17. What is a seawall and how are they designed to protect beachfront property?

Q18. Is a seawall a positive or negative fix and why?

Q19. What is a groin in terms of oceanography? How are they designed to protect the beach?

Q20. If there is an easterly wind (blowing from east to west) with a groin running north to south from the beach, where will deposition and erosion take place? Finish the sketch in figure 7.5.

Figure 7.5

Beach

Ocean

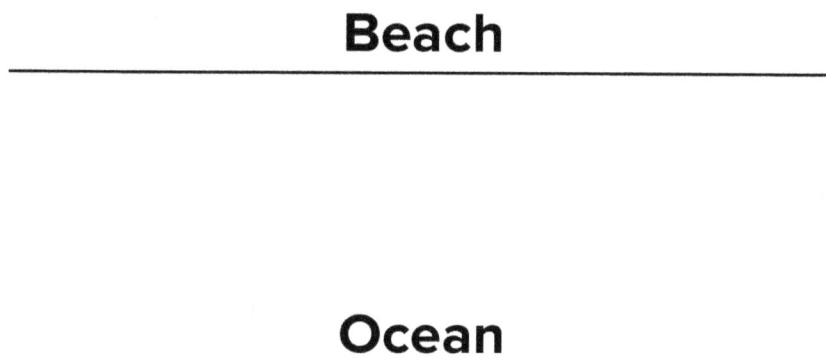

Q21. What is a jetty? How are they designed to protect the beach?

Q22. How is a jetty different than a groin in terms of erosion and deposition of the beach around the jetty?

Q23. What eventually happens to a beach if a breakwater is built?

8 Productivity

Purpose

To understand the connection between temperature, ocean circulation, and productivity in ocean waters. To understand more about the plankton that are associated with oceanic productivity.

Materials

- ❑ Microscope slides: foraminifera, copepod, radiolarian, and diatom
- ❑ Colored Pencils

Background

For this lab, you will use satellite images showing primary productivity in various locations on Earth (https://earthobservatory.nasa.gov/images/6735/a-world-of-chlorophyll or scan QR code). The colors in the pictures are related to chlorophyll and not temperature, so make sure you don't interpret temperature by using the colors seen on the map. The colors' meaning is described in table 8.1.

Table 8.1

Color	Productivity
Red and Yellow	High
Light Blue and Green	Moderate
Dark Blue and Green	Small
Black	No Signal

Part 1: Productivity and Rivers

The Orinoco River is in the northeastern part of South America. Use the satellite picture to answer the associated questions (figure 8.1).

Figure 8.1

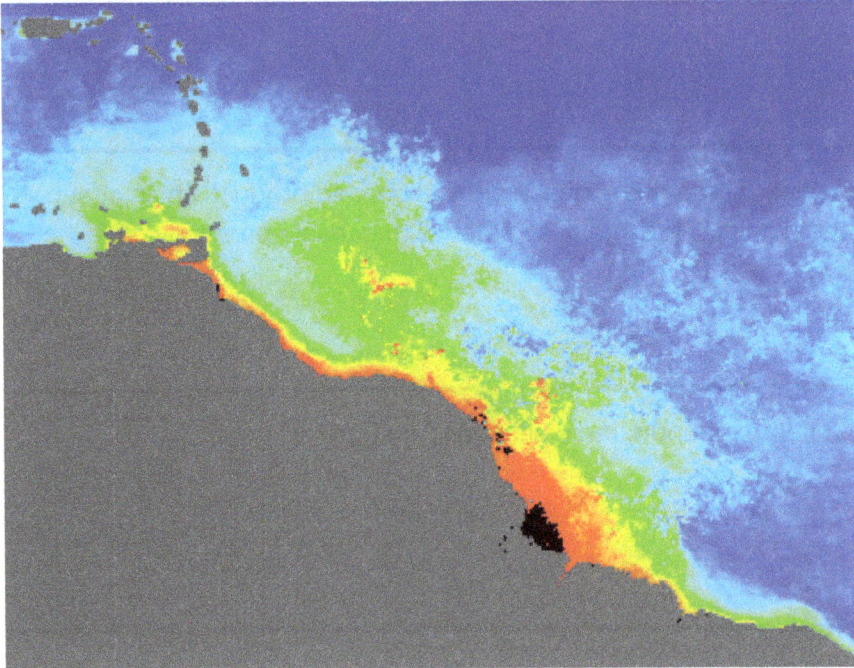

Q1. What causes productivity to be high in this particular location?

Q2. Sketch the ocean currents on the picture and label them. Look north and south of the black area, where the river is located. Why are nutrients more abundant and spread out north or the Orinoco River?

Notice that productivity is high inside the Gulf of Carpentaria and Sea of Arafura between New Guinea and Australia (figure 8.2).

Figure 8.2

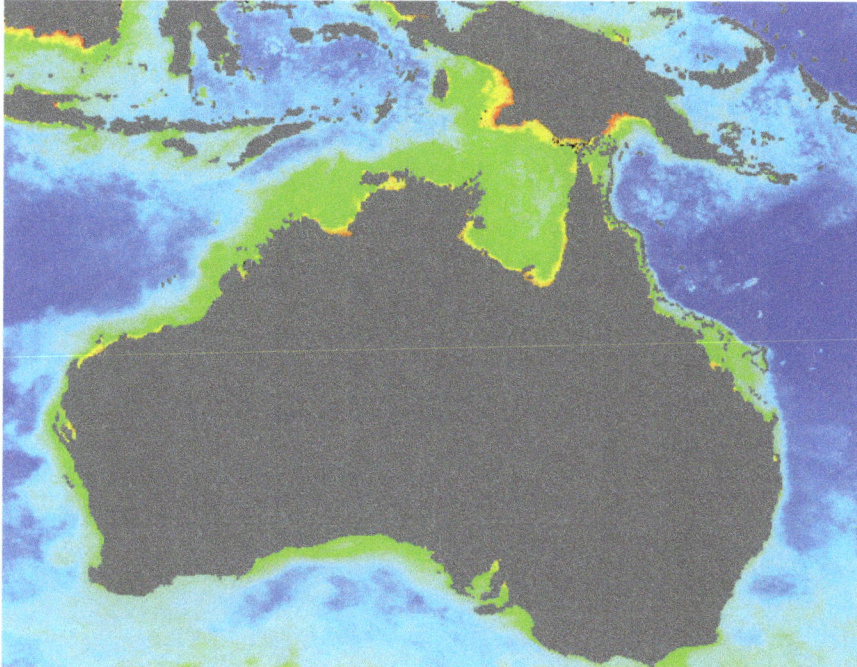

Q3. Why would productivity be high in this area?

Q4. Sketch any associated ocean currents on the image. How could the ocean currents maintain the high productivity in the Sea of Arafura and prevent the productivity from extending further into the Pacific Ocean (the right side of the picture)?

Part 2: California's Coast

This map shows productivity on the west coast of the United States (figure 8.3).

Figure 8.3

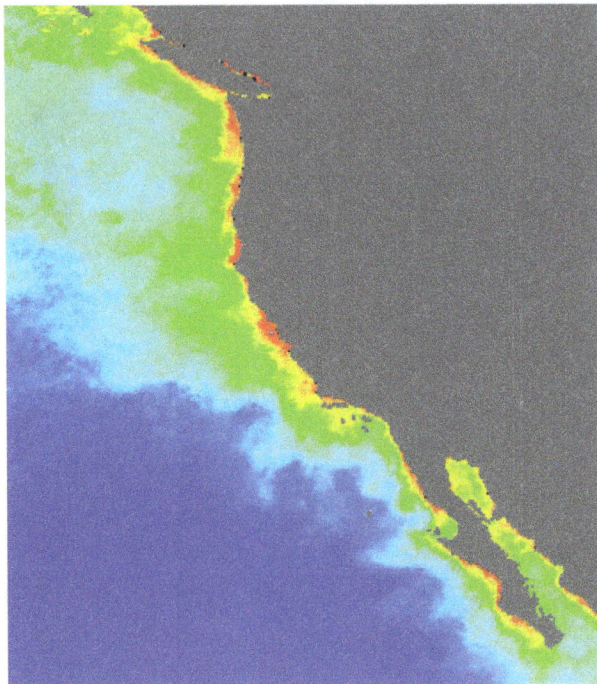

Q5. Sketch the California Current on the map. Sketch the direction of the Ekman Transport along the California Current that you previously sketched.

Q6. What is the general temperature of the California Current? How could that help support plankton production?

Q7. What process does the related Ekman transport create along the California Coast? How could that process help support plankton production? Give two reasons.

Part 3: Equatorial Productivity

The Congo River is on the central west coast of Africa, about 5° south of the equator. Use the satellite picture to answer the associated questions (figure 8.4).

Figure 8.4

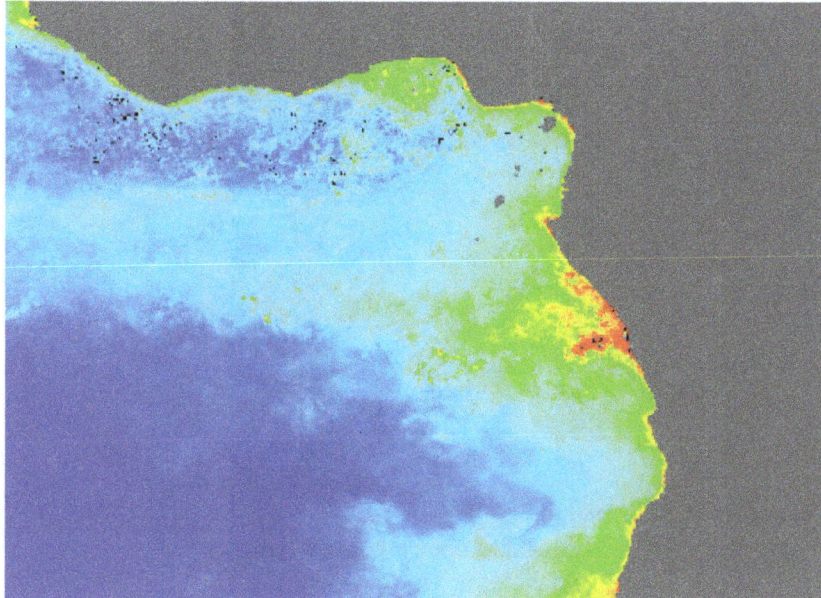

Q8. Sketch and label the directions of the trade winds. Then, using a dotted line, sketch the associated ocean currents. Finally, add the direction of Ekman Transport along the ocean currents.

Q9. How is upwelling created by the motion of the ocean currents here? Why would this increase productivity? Give two answers.

Part 4: Plankton

Use the microscope to view different plankton samples. In table 8.2, sketch and name the sample, then circle the type of plankton below; whether it is a cold or warm water species; and whether it is made of calcite or silica. I'm not grading you on the awesomeness of your drawing, so do your best.

Table 8.2

Sample A: _____ (Zooplankton or Phytoplankton) (Cold water or Warm water) (Calcite or Silica)	Sample B: _____ (Zooplankton or Phytoplankton) (Cold water or Warm water) (Calcite or Silica)
Sample C: _____ (Zooplankton or Phytoplankton) (Cold water or Warm water) (Calcite or Silica)	Sample D: _____ (Zooplankton or Phytoplankton) (Cold water or Warm water) (Calcite or Silica)

Q10. At what general depths would you expect to see calcite type plankton? Why?

Q11. At what general depths would you expect to see silica type plankton? Why?

Part 5: Global Productivity

Use the satellite image in figure 8.5 to answer the questions about productivity at different latitudes.

Figure 8.5

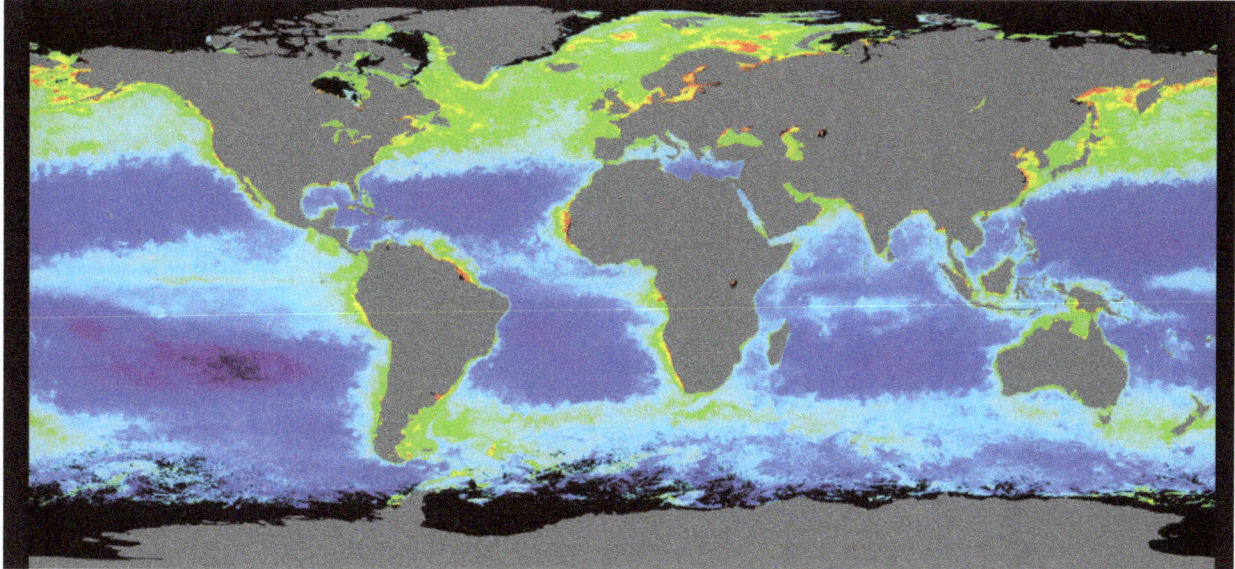

Q12. What time of year do you think this satellite picture was taken (use months and not seasons as seasons are different depending on hemisphere)? How can you tell?

Q13. Where is productivity the lowest? Why?

Q14. How can you locate the equator on this map? Why does this occur?

9   # Climate Change and Ocean Acidification

Purpose

To understand the fundamentals of climate change and how it can affect the ocean.

Materials

- ❑ Calculator
- ❑ Litmus paper
- ❑ 250 mL beakers
- ❑ Forceps
- ❑ Lemon juice
- ❑ Window cleaner
- ❑ White vinegar
- ❑ Alkaseltzer

- ❑ HCl
- ❑ Sample of Coquina
- ❑ Distilled water
- ❑ Seawater
- ❑ Bromothymol Blue
- ❑ Pipette
- ❑ Straw

Part 1: Carbon Dioxide in the Atmosphere

The Earth receives about 240 watts/m² of heat from the Sun every second, which is roughly the same as the amount of heat that is emitted by the Earth into space. The sun emits visible light (UVA and UVB rays) that the Earth absorbs and converts to infrared rays. These are emitted by the surface of the Earth back into space.

The amount of heat emitted by a "black body" (an ideal radiator and absorber of energy at all electro-magnetic wavelengths) is provided in the graph in figure 9.1. The data is from laboratory experiments where an object is heated and its temperature and the heat it emits are both measured. This helps us see approximately how much heat the Earth emits.

Figure 9.1

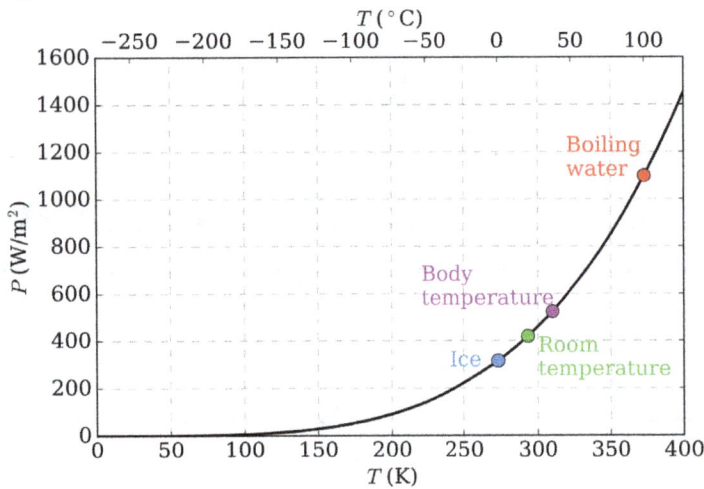

Q1. If each square meter of Earth's surface receives 240 watts/m² of heat from the sun, what should the hypothetical temperature of Earth be in Celsius (°C)?

Q2. Based on your own personal observations of living on Earth, is the average temperature similar to what you estimated above? In our area, what do you believe the average temperature is (use centigrade)?

Q3. If your estimate in question 1 was incorrect, what causes the Earth to be a different average temperature than explained by laboratory models? Why? *Hint: Think about atmospheric composition.*

Q4. What are the main greenhouse gases in the atmosphere? What happens as these gases increase in the atmosphere?

Figure 9.2

Global atmospheric carbon dioxide compared to annual emissions (1751-2022)

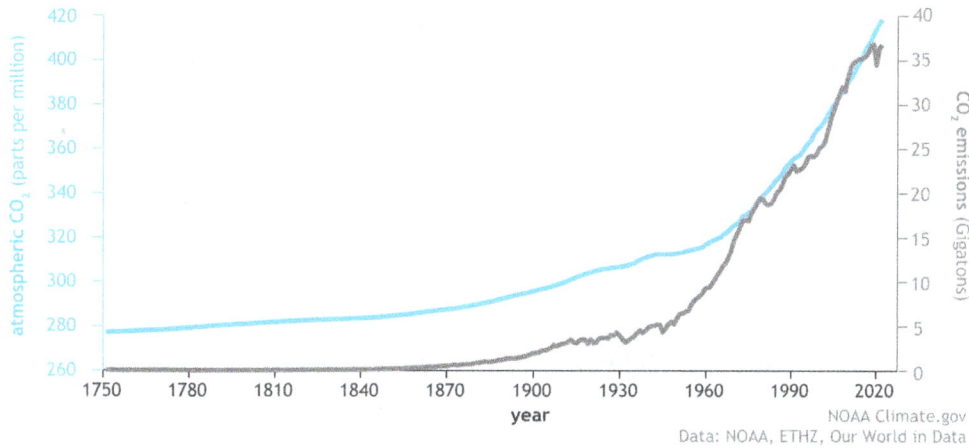

NOAA Climate.gov
Data: NOAA, ETHZ, Our World in Data

Q5. Using the graph in figure 9.2, what was the concentration (in parts per million) of CO_2 in the atmosphere between the years of 1750–1850 (100 yrs)?

Q6. How does the amount of CO2 change by the year 2020? What is the percent increase in CO2 since 1870? Show your work for the following equation:

((Final – Original) / Original) * 100 = ?

Q7. What occurred around the 1800s that would have increased the amount of CO_2 present in the atmosphere? Why would this increase the amount of CO_2 in the atmosphere?

Q8. What happens to the temperature on Earth as the CO_2 in the atmosphere increases? Why?

Q9. If direct measurements of CO_2 in the atmosphere were not taken until 1958, how do you think we got the estimates of atmospheric CO_2 for the graph? Do you think these measurements are a good approximation of atmospheric conditions of the past?

Part 2: pH of fluids

Materials

- Beaker
- Litmus paper
- 250 mL beakers
- Forceps

- Lemon juice
- Window cleaner
- White vinegar
- Alkaseltzer

- HCl
- Sample of coquina
- Distilled water
- Seawater

Background

The oceans currently behave as carbon sinks because they absorb approximately 25–30% of the carbon dioxide put into the atmosphere by human activities. To understand how that affects the oceans, we first need to understand pH and the pH of seawater.

The concept of pH is best understood as a scale from 0–14, with lower numbers representing acid substances and higher numbers representing alkaline substances. Table 9.1 shows this information.

Table 9.1

Substance	pH	Ions
Acid	0–7 (or less than 0)	More H^+
Neutral	7	Equal H^+ and ^-OH
Base (Alkaline)	8–14	More ^-OH

Testing the pH of a substance can be done easily using litmus papers, which show a colorful reaction to different pH levels. For reference, mixtures of acid and basic substances produce solutions closer to the neutral range (e.g., Acid + Base = Neutral, or $H^+ + {}^-OH = H_2O$, that is why pure water is considered neutral pH).

Procedure

1. Estimate the pH for the following solutions by dipping litmus paper into the solutions using your tweezers (please throw your litmus papers away and rinse your tweezers after each sample). Then, record your data in table 9.2.

Table 9.2

Fluid	pH (use your best estimate)	Acid? Base? Neutral?
Lemon Juice		
Baking Soda in Water		
White Vinegar		
Alkaseltzer		

Q10. Look up the pH of the solutions using the internet. Were your estimates correct?

2. Make sure you rinse out your beakers really well before adding your water samples, and rinse your tweezers.
3. For the next measurements, you are going to use litmus paper in the same way as above, recording your answers in table 9.3.
4. Once you have measured the initial pH of the two water samples add a couple of drops of the hydrochloric acid (HCl) provided and stir well.
5. Retest their pH with litmus paper. First, estimate what their pH will be after acid is added and record your answers in table 9.3. Then, add the acid, test the solution, and record the data.

Table 9.3

Water type	Estimate pH	pH after adding acid (HCl)
Freshwater		
Saltwater		

Q11. Which sample is more acidic before adding the acid? After adding the acid? Do your results make sense?

Q12. Seawater has more ⁻OH in it than regular water. What should happen to the acid you add to seawater based on the important information in the background section of part 2?

Q13. What will happen to the ocean if it absorbs too much acid? Will its pH change?

Q14. Find a sample of coquina where the lab materials are. This sample is made of Calcite ($CaCO_3$). Describe this rock sample. What do you think composes this rock?

Q15. Take the acid (HCl) provided and put a couple drops on the rock. What happens as you add acid to this sample? What is the acid doing to the rock sample?

Q16. How do you think acid affects animals who make shells out of calcite? Would animal life prefer more acidic seawater or more basic seawater?

Part 3: Carbon Dioxide and pH of water

Materials

- Beaker
- Distilled water
- Phenol Red
- Pipette
- Straw

Background

Phenol Red is an acid-base indicator. It is red in neutral solutions and turns to orange as the solution becomes more acidic and eventually yellow as the pH becomes even more acidic.

Procedure

1. Fill a beaker with 150mL distilled water.
2. Add several drops of phenol red.

Q17. What is the color of your solution when you add phenol red to distilled water? Is it acidic, neutral, or basic?

3. Insert a straw into your solution and blow bubbles slowly into the solution. **Do not suck the solution into your mouth.**
4. Repeat this process 5 times.

Q18. What happened to the color of your solution after every breath? How did the pH of your solution change?

Q19. What are you exhaling that could be affecting the pH of distilled water?

Q20. Other than respiration, what are some other ways that more CO_2 can be added to the atmosphere?

Q21. In part 1, you described the change in CO_2 in the atmosphere. How do you think this change in the atmosphere has changed the pH of seawater on Earth?